Isles, People..

Athalie Ducrotoy

– A family in olde Seychelles

1780 - 1995

Beautiful Isles, Beautiful People.

Athalie Ducrotoy

First published 1995

ISBN 0 9526556 0 8

Published by
Rawlings Publications,
P.O.Box 175, Kent CT14 6GD

ACKNOWLEDGEMENTS.

His Excellency, President René of Seychelles, for his permission to use the Official Archives.

British Library, London.
Reuters, London. (December 22 1981)
Ministry of Defence, Army Medal Office, Worcestershire.
Danielle de St Jorre, Minister of Foreign Affairs, Victoria, Mahé.
Guy Lionnet, Victoria, Mahé, Seychelles.
Director of Archives, Pointe Conan, Mahé, Seychelles.
Jackson, Flavie. Seychelles Bulletin (May 4.1974) Mahé.
Mayor of Vignacourt, France, for the use of documents.
McMillan Museum, Nairobi, Kenya.
Mr McGaw (now retired) previous Director of Archives, Mahé, for his continuous help over the years.
Rita Jorre de St Jorre, Australia for the use of photographs.
Nancy and Philip Savy, Australia for various documents.
Monique Deray in Vignacourt.

FRONT COVER: By permission of the British Library, London. Map of Seychelles 1776.69490 (1)

CONTENTS.

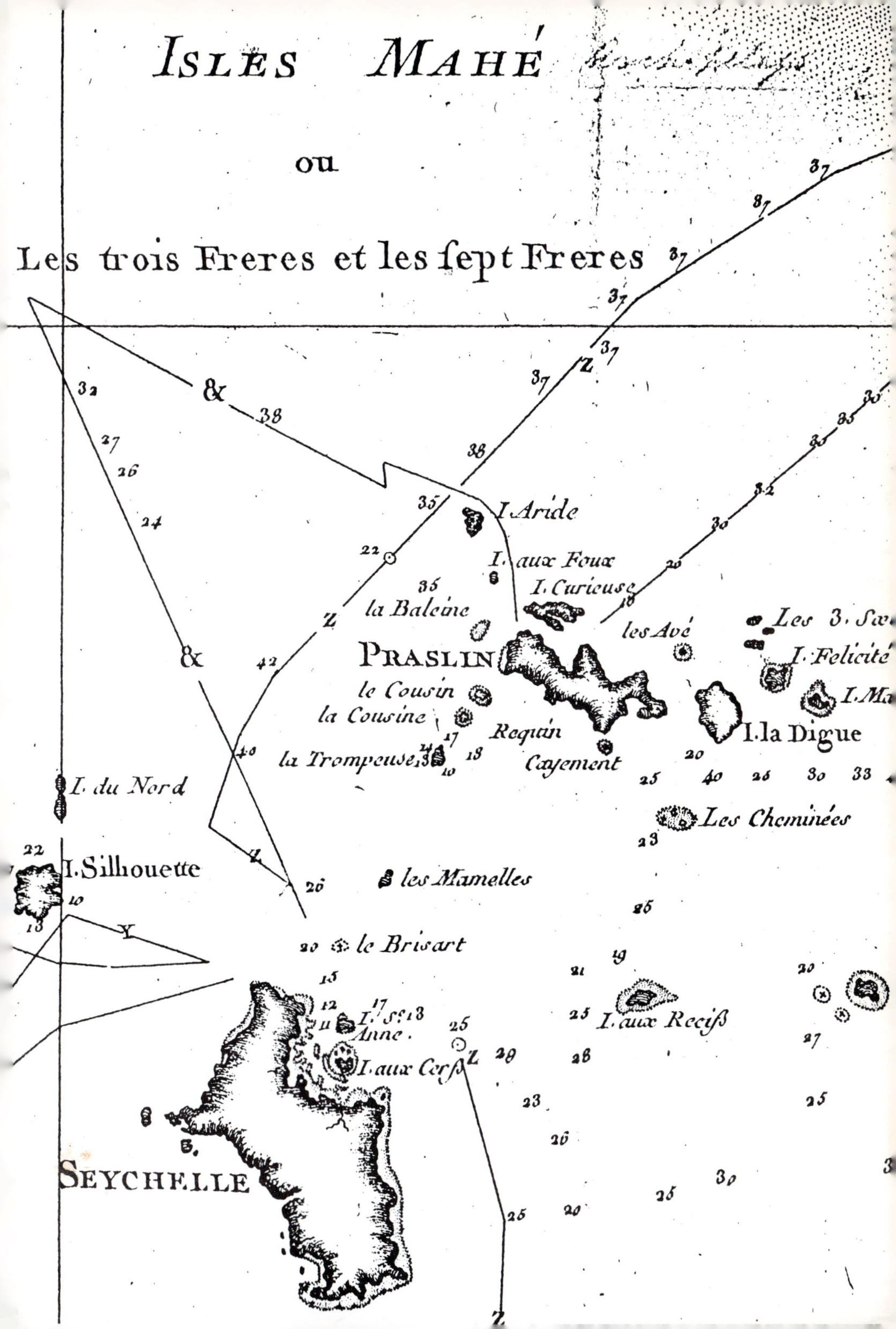
Isles Mahé
ou
Les trois Freres et les sept Freres
I. Aride
I. aux Foux
I. Curieuse
la Baleine
Praslin
les Avé
I. Felicité
I. la Digue
le Cousin
la Cousine
Requin
la Trompeuse
Cayement
Les Cheminées
I. du Nord
I. Sillhouette
les Mamelles
le Brisart
I. Ste Anne.
I. aux Cerfs
I. aux Recifs
Seychelle

THE LINEAGE OF THE TWO FAMILIES.

An easy reference to the lineage of the Ducrotoy and Savy families.

Ducrotoy 1740	Charles Savy of Bourbon and Seychelles 1730
Antoine Ducrotoy	Savy of Seychelles
Florent Ducrotoy of France	Victor Francois Savy of Seychelles
Joseph Ducrotoy of France	Vildoric Victorin Savy of Seychelles
Philibert Olivier Julefy Ducrotoy France & Seychelles	Marie Elisca Savy of St Anne, Seychelles
Philibert Olivier \Julefy Ducrotoy = 1896 - 1954	Vivienne Maude Savy 1906 - 1974

Dedication

Dedicated to

Maude Vivienne Savy 1906 - 1974

whose tales of her island home of Saint Anne on Mahé gave me the inspiration to research and record the following pages.

This is a passport photograph of Maud taken in 1923 before leaving Seychelles for Kenya to meet her future husband, Philibert Ducrotoy, in Kenya. The official stamp impression can just be seen.

HISTORY OF SEYCHELLES IN RELATION TO THE FAMILY.

200 - 300BC Islands visited by Polynesians.
1501 - Visited by Portuguese.
1609 - British Expedition, the East India Company.
1742 - French first landed.
1756 - Stone of Possession set up by French.
1760 - Pirates and castaways.....
1770 - French Colonists sent from Mauritius, slaves from Africa, settlement on St Anne Island.
1780 - Adelaide born (Savy ancestor).
1794 - British captured Seychelles.
1798 - Victor Francis Savy born Seychelles.
1808 - Bidale Lalanne born Mahé.
1814 - British won Seychelles by Treaty of Paris.
1827 - French Commandant Chevalier Queau de Quincey died.
1835 - Slavery abolished. Jaques Edmond born Seychelles.
1837 - Henrietta Lebon born Seychelles.
1843 - Vildoric Victorin Savy born.
1855 - Jaques Edmond and Henrietta Lebon married
1881 - General Gordon inspected the Isles and drew up Coat of Arms.
1882 - .Philibert Olivier Julefy Ducrotoy left France.
1885 - Philibert Ducrotoy landed in Seychelles.
1906 - Maud Vivienne Savy born on St Anne Island, Seychelles.
1914 - Philibert Ducrotoy joins East African Mounted Rifles, Kenya.
1915 - Philibert Ducrotoy wounded near Seki waterhole, Kenya.
1926 - First electric lights installed on Mahe.
1929 - Albert Reginald Ducrotoy born Mahé, Seychelles.
1930 - René Ducrotoy born Mombasa.
1935.-.President Albert France René born in Seychelles.
1935 - First automobile on Mahe, Seychelles.
1976 - Seychelles granted Independence from Britain.

INTRODUCTION.

The people of Vignacourt in France still remain in the quiet village to this day as many generations of ancestors have before them. However one of their kin hastily left the village one night while they slept and his adventures subsequently led him to Seychelles.

The Ducrotoy generations that followed have reason to be thankful for grandpapa's bad temper for without it the family name would have become extinct. As a youth grandpapa Ducrotoy quarrelled violently with his stepfather and fearing the consequences rapidly left his family to join the French Foreign Legion at Marseilles. From there he was transferred to Sidi in Algeria. After a formidable training he was shipped off to Indo-China but Philibert had different plans.

He deserted the ship at Mauritius and found his way to Seychelles where he settled and started a coach-building business. The decision to settle and raise a family in Seychelles, never to return to France, saved his life. All his male relatives were killed in that dreadful battle of the Somme, France during the 1st World War many years later and but for his flight he would have fallen with them.

In view of the fact that he had deserted from the Foreign Legion, an offence for which he would have been shot, he was unable to join up to help save his country in the Great War. Instead he enlisted with the East African Mounted Rifles in Nairobi, Kenya. He narrowly escaped death when his horse was shot by Germans at Seki waterhole, and he laid throughout the night, injured, with a dead horse on top of him as lions ate the corpses of his companions around him.

Philibert's son married a descendant of Charles Savy the notorious Mahė slave trader. Some of the family descendants remain in Seychelles today though many are scattered throughout the world yet they have two things in common. The first is a universal kinship towards their compatriots and affinity for the Seychelles. The second, they were invariably raised on these enthralling stories of olde Seychelles some of which are recorded here. This is the story of just one family but there are many similar stories waiting to be told.

A.D.

On The Road to Saint Louis, Mahe, Seychelles.
1906

DUCROTOY FAMILY TREE
Vignacourt 1866

Louis Barthelomy Ducrotoy
= Celestine Domersal

Antoine Ducrotoy
1800

Louis Ducrotoy
12 April 1812

Desire
1805
(retailer)

Florent Ducrotoy
b. 7 Jan 1811
= Judith Pruvot

Alcide Ducrotoy
23 Feb 1837
(woodcutter)

Edward
b.1841

Joseph Ducrotoy
b. 15 July 1835 6pm
died Amiens 18 Dec 1865
= Irma Thuillier >

Joseph Alfred Ducrotoy
25 Oct 1861 Vignacourt
= Marie Aurelie-Angelie

Philibert Olivier Julefy Ducrotoy
13 August 1866, Vignacourt
= Marie Julia Edmond of Mahé,
Seychelles at 8am,24 Dec 1891.

See separate family tree for
family remaining in France.

See family tree for descendents
in Seychelles.

THE WEDDING OF HENRIETTE DUCROTOY.
2 August 1911 Mahé, Seychelles.

The marriage of Henriette Ducrotoy and Henry Ernestine of Mahé, Seychelles. Philibert Ducrotoy (senior) with moustache and top hat. Rene's father, also Philibert, is the taller boy in the white suit centre of group then aged 15 years. The smaller boy is Joseph Ducrotoy later killed in a carriage accident.

BEAUTIFUL ISLES, BEAUTIFUL PEOPLE...

CHAPTER 1

-Far from the islands of Seychelles a child was being born on Rue de Labas in the village of Vignacourt in France. It was nine o'clock in the morning on the thirteenth day of August 1866 when the elderly midwife delivered the posthumous son of Joseph Ducrotoy. Joseph had died many months before in Amiens the nearest market town to Vignacourt. It would have been devastating for his widow to learn soon after the death that she was expecting a second child. Irma returned to her parents Olivier and Virginie Thuillier for the delivery. The child was named Philibert Olivier Julefy Ducrotoy and born to travel far from this small French village. He had an older brother, Joseph, born five years before and he was to generate the family remaining in the village after Philibert had set forth to foreign shores.

The two boys continued to live in their maternal grandparent's home and attended the village school in Vignacourt. Some years later their mother married again and this led to Philibert's eventual departure from the village and France. It was well known that Philibert and his stepfather clashed on many an occasion but this was no reflection on Mr Pécourt as Philibert displayed, on occasion, shortness of temper most of his life. What followed led to Irma's youngest son leaving home at the age of sixteen never to return. This momentary decision probably saved his life as most of his contemporaries and male relatives died at the great Battle of the Somme, France. Philibert and Prophet Pécourt had a violent row during which Philibert hit his stepfather severely. In fear of being arrested he left town that night and made his way south to Marseilles. Not being able to gain employment without papers or money he made his way to Fort Saint Jean at the old harbour of Marseilles. With a group of others he enlisted in the French Foreign Legion. The training was formidable but there was no going back. Eventually grandpapa was sent to the Legion fort in Algeria and after marching in the desert heat with his pack upon his back finally entered the gates of Sidi-bel-Abbes. As they approached the intense heat shimmered off the dry walls of the fort and lizards ran and sheltered in the cracks of the stone. Each day the enlisted men were sent out, with a large pack on their shoulders to walk many miles into the Sahara as a test of endurance and strength. Their backs were loaded with weapons, spades, food and firewood. Philibert was accustomed to a temperate climate and the temperature of over one hundred degrees became extremely difficult but he survived. It was here that he had tattoos designed on his body which he was to carry for the rest of his life. Across his thighs he had the mark of the Legionnaire inscribed, "Marche ou Crève" (March or Die). 'Marche' on the right thigh and 'ou crève' on the left. Across his chest was also written "Enfant du Malheure", (a lad of misfortune). In the light of his subsequent escape from the Legion this was a foolhardy move for this identified him as a Legionnaire by all authorities. There is no question that he would have been handed back by any military division and there would have been grave consequences. However these were early days...

In 1883 the French conquered Indo-China and that year the first battalion left Sidi for the Far east. other battalions followed to replace those having fallen in battle or overcome by tropical diseases.

Steamer to Indo-China.

Along with a contingent of young Legionnaires Philibert was shipped off around the Cape of Good Hope to Indo-China for military duty. There were alarming stories filtering back to them about the privations they might encounter so Philibert devised a plan. This plan was, with several friends, to desert as soon as the ship came into port. No-one knew of course which port that would be. During the long sea journey there was much speculation as to the difficulties they might meet.

The ship could land at Reunion which was French so there was little chance of escape there. The great worry was the prospect of arrival at Madagascar which was then in turmoil and just as dangerous and feared as Indo-China. If landfall should be the coast of Africa then Philibert and his friends could not speak the language. It was some relief therefore when one morning they awoke to find themselves

PHILIBERT OLIVIER JULEFY DUCROTOY
1866 - 1924

Born in Vignacourt, France, married in Seychelles and died in Kenya. He fought in World War 1 in the East African Mounted Rifles and East African Medical Corps in Kenya. This photograph was taken in 1921 when he was fifty five years old and several years before his death.

at Mauritius. Philibert and his three friends lost no time and after jumping ship hid with local families on the island. The officers of the Legion failed in their extensive search...

A schooner was leaving port and having been reassured that it was not landing at Madagascar the men set off as deck hands bound for Seychelles. The year was 1888.

Philibert had found the perfect place to settle. A picturesque island, English yet French speaking, and quiet and secluded enough to escape recapture. Philibert would meet with the friends he arrived with and sit long into the night singing the old French songs and tearfully remembering their training days and France. Before long he became engaged in various enterprises which gained him the attention of one Jacques Edmond.

The Edmond family were educated and Jacques considered this ambitious and industrious Frenchman a suitable catch for his only daughter. Marie Josephe Julia Edmond accepted the idea so the wedding was planned...

The coaching business...

Philibert Olivier Julefy Ducrotoy and Marie Josephe Julia Edmond were married at eight o'clock in the morning on the 24th December 1891. With the consent of the bride's parents who were present at the Civil ceremony, the witnesses were Elysee Savy an Attorney-at-Law and Charles Edmond, Government servant her brother. The religious ceremony was performed following the official one. At a time when few people were educated, in particular women, Julia signed the Register with perfect handwriting, as shown below.

Signature of the married parties.

Philibert Ducrotoy

Julia Edmond

Julia's father helped Philibert establish a coaching and blacksmith business. He had wagons and carts to hire out or use for deliveries and closed-in coaches imported from France which were used for funerals. These were black with black plumes and the black carriage drivers looked extremely smart in their top hats. Two examples of the carriages can be seen on page 12 taken at a family wedding in 1906. The smaller, open seater, a pousse, was used for church on Sunday.

Eleven children.

Between 1892 and 1906 Philibert and Julia produced eleven children all born in Seychelles. Two of the children, sons, one died at birth and the other at three years of age. The remaining children were; Bella, Henriette, Philibert (who later married Maud Savy and became René's mother), Edma, Marie Jeanne, Joseph, Marie Emily Paquerette, Octavia and Julia.

It is uncertain to which school the children attended. On Mahe were St Pauls', Bel Ombre, Glacis, St Louis and St Josephs. With the exception of St Josephs' and St Louis which were free, school fees in 1905 were 2 to 3 rupees a month depending on the grade. All those surviving infancy, except for Joseph who was killed at the age of sixteen by a bolting horse, eventually married and are chronicled towards the end of this book.

Grandpapa proudly declared that he never worked for anyone but himself. His coaching business continued and prospered. Hire for a hackney carriage was 20 rupees a year and a hackney driver 2 rupees.

First World War and East African Mounted Rifles.

Philibert felt the urge to enlist, initially, not with any patriotic tendencies towards Britain, but sympathy for his beloved France. News reached them in Mahé that France had fallen and his male relatives killed on the Somme and he rushed to sign up. The French Consul, however, reminded him that he was a deserter from the French Foreign Legion and as such he would be shot. The kindly Consul suggested he sail to Kenya and join the British Forces there since they were fighting the same cause. This decision decided the family's future and their consequent departure from the shores of Seychelles.

Enlistment.

Those arriving from Mombasa by ship told stories of the men in Kenya marching to the towns to sign up and Philibert announced to the family that he was off to Africa to take up arms. On arrival he registered with the East African Mounted Rifles at Tsavo, a largely amateur force of eager volunteers formed through necessity. The troops in Kenya, the King's African Rifles, were deployed but owing to huge losses both from war casualties and malaria every man was welcomed, any age any skill...

Philibert joined an assorted bunch with Squadron E on the 11th February 1915 and set off with a variety of horses and mules camouflaged as zebras with black and white paint. The troops encountered many difficulties. Food had to be carried, not only for the men but for the horses who also needed constant water. Besides, Philibert was accustomed to being waited on by servants, a wife and seven daughters. Now he was having to light fires to cook meagre rations so life was difficult.

He recounted several stories over the years about that period as follows. There were British Forces in Tanzania fighting in a kapok plantation but they were seriously outnumbered by German troops. In despair, the British kept firing knowing that time and ammunition was limited. By a fluke one of the British bullets dislodged a wild beehive settled near the trunk of a handsome kapok tree behind the German line. The German troopers enveloped by the increasing swarm of angry bees screamed and rapidly retreated...

The second story came much later. A German returned after the War and was sipping tea with a plantation owner and he was telling how, during the war, he, with a group of others, had planted a mine beneath a railway bridge. They waited for most of the night for the Voi train to pass without success. The plantation owner listened quietly to the story and said, “If you *had* been successful we would not be having this conversation. I was aboard that train.”

It seems that the Voi train had been delayed through the night and the officer had decided to wait till first light. In the morning before setting off scouts were sent forth to check and found the mine attached to the bridge...

The night spent with a dead horse.

History and grandpapa's stories differ slightly at this point. Records suggest that Philibert's squadron were on the Kajiado to Longido road and that he was injured at the Seki water-hole. Philibert always told the family he was wounded at Taveta. Taveta and Seki are quite a distance apart so this is a little uncertain. It may be that he was carried to a field hospital at Taveta.

It was the 13th May 1915 when his horse was shot by the enemy, the shot passing through the horse and into his ankle causing a serious wound and the horse fell dead with Philibert trapped beneath it. After he recovered from the shock he tried to shift but due to the extreme pain of his injured ankle and the weight of the horse movement was impossible. He was aware of others wounded around him and he could hear a man delirious and talking in English. Now Philibert could only understand a few words of English but he heard the man saying,"Oh, my poor wife. Oh, my poor wife." Over and over he cried out until he eventually went silent. He did not survive...

The predicament in which grandpapa found himself in slowly dawned. He could call to the German troops who were still in the vicinity and risk being shot. It was always understood that neither side could take prisoners due to the difficulties in transporting and feeding. He could stay where he was with loss of circulation to his leg and risk losing it. There were hungry lions roaring in the distance and dead flesh about him. Philibert was unsure whether help would arrive in time or even if assistance would come at all.

During the evening three lions came by as Philibert laid helpless. They sniffed the air and walked on. After dark he heard them gnawing at flesh some yards away. He considered the situation and reasoned that their was sufficient horseflesh to satisfy the lions before they would choose a man fully clothed in his Army kit. That is, if help came in time.

He was found the following day and transported rather painfully by the Red Cross back to the Army hospital by rickshaw.

Many weeks of recuperation followed and the records show that Philibert stayed in the hospital working. This was unknown to the family until Army records were released fifty years later. Philibert was considered an excellent horse 'doctor' by everyone who knew him. It is quite possible, not being fit for active duty, he was kept on at the field hospital to help the wounded. War records in London state that he worked in the East African Medical Services until the 6th December 1915 when he was discharged with a small pension for life.

The family sails to Kenya.

Following his discharge from the Army it was still wartime of course and he considered that there were possibilities for the family in Kenya. He returned to Mombassa and wrote to the family in Seychelles. he instructed them to pack everything and bring it all to Kenya, and he meant everything. This was no easy task. There were masses of tools, forge blower, spanners, carpentry tools and even the blacksmith's anvil came. All the equipment to construct and repair coaches not to mention furniture and clothes and the remaining nine children and servants.

Amongst the servants there was a young boy that the Ducrotoy family were fond of, entrusted to their care by the parents in Seychelles. This family hoped the boy would prosper and return eventually to Seychelles. Years later as grandpapa lay dying he made grandmama promise to take care of the lad. Unfortunately the servant died of malaria a few years later.

Coffee over the Customs' Officer.

Now Philibert was an impatient man. He had been working at the stables owned by Lord and Lady Colville and was anxious to have his tools so he could continue an important job. He waited and waited in the hot sun and eventually the ship arrived with Julia and all the children. He was most anxious to clear the baggage and commence the journey back before dark. The Customs' Officer took his time and besides he did not understand French. Philibert became more and more heated and started pacing about. Grandmama knew it wouldn't be long before there was a explosion. The officer asked to see one box then the other, using a jemmy to lever open a wooden crate full of tools, then started on the suitcases... He came across a tin of ground coffee that Julia had prepared before leaving and removing the lid asked, "What is this?" Philibert by this time was not only heated but boiling with fury, nevertheless, he ignored the officer. The question was asked again, but in a louder voice this time, *"What is this?"* With that Philibert stalked over, threw up his fist hitting the bottom of the tin of coffee which spewed its contents all over the Customs' officer's white starched uniform and down his face. The enraged officer called the wharf police and was about to have him arrested when Lady Colville arrived and took the situation in hand. She led the officer to one side and told him, "Please forgive this man, he is a wounded soldier straight from the front-line. He is tired and worried about his family, besides he is known to me..." Lady Colville being in charge of Soldier's Welfare spoke with authority. Slowly and reluctantly the officer finally agreed to let the matter drop and completed the forms promptly...

Our father...

In Mombassa there was a soap manufacturer, Mr Dubois, who drove into town each day in his box-body car wearing his pith helmet. Looking very distinguished he would step out of his car in his long black coat and sporting a long white beard. Now, in Seychelles at that time every priest had a beard so the Ducrotoy girls politely greeted Mr Dubois each day with, "Good morning, Father". Very soon all the other shop assistants were doing the same much to Mr Dubois's embarrassment...

Grandpapa supplied Mr Dubois with pork fat and sausages from his farm and one day he was addressed, "Mr Ducrotoy. Please tell your daughters not to call me 'father'." It seems that his neighbours who knew he was unmarried wondered how many children he had...

CHAPTER 2

WHEN VINEGAR TURNED TO WINE.

Grandpapa always made his own wine and beer with sugar cane, fruits such as pineapple, oranges, even potatoes but the making of such was a monopoly of the authorities and as such, illegal. This, however, in no respect deterred Grandpapa. His family had been brewing for centuries in France through several Revolutions and he couldn't see a reason to stop, enabling to British importers to prosper, so production carried on as usual.

A neighbour, who Grandpapa detested and called 'nez crochu' (hooked nose), informed the police and during the raid of the farm many large wooden barrels were removed for analysis and evidence. In 1919 laboratories for testing were few and far between, besides, the train services most infrequent, so the barrels sat in the Mombasa Police Station in the heat of the summer. As the Officer sat preparing the documents for despatch the brew merrily bubbled away and fermented until it spoiled.

The evidence was duly forwarded on the Nairobi train for detailed analysis from the scientist there so the Court case could be prepared.

Philibert turned up on the day of the Court session in his best suit. A translator was brought in to translate from English to French which did nothing to nullify the general confusion.

The Judge sat sternly on the Bench and addressed Philibert, "Could you please tell me the contents of the barrels removed from your property?"

Philibert, being proficient in the skill of making the illegal liquor and had predicted the outcome confidently replied, "Vinegar!"

The Judge, trying to maintain his composure, asked the Court Prosecutor to hand him the scientific report from Nairobi and agreed, "Yes, it is true. The laboratory report states that the liquid is vinegar." He glared at the embarrassed Police Officer who had brought the prosecution.

Almost as an afterthought the Judge questioned, "Why do you make so much vinegar?"

Philibert responded meekly with his hat clutched in his hands, "I preserve meat in vinegar, cabbage, fruits, eggs, everything...., and your Honour, I have a large family to support."

The British Judge seemed bemused but not surprised at any delicacies the French would consume, and pronounced, "Case dismissed. The matter is finished."

With that, Grandpapa straightened and with increasing confidence he shouted out in French, "C'est pas tout! Je veux le contenant a le contenu." (It is not finished! I want the container and the contents.")

The precious barrels of vinegar were eventually loaded onto the horse and cart and returned to the house. That evening Philibert had the servant bring bags of sugar and this was added to the containers and by the following morning the brew was fermenting with vigour once more.

A grand party ensued when the fermentation was over and all agreed that it was the strongest brew so far. Grandpapa raised his glass to the neighbour who had reported him, "Come on, Nez Crochu, come and have a drink with me..."

Grandpapa's daughter-in-law Maud Savy came by ship from Seychelles to marry his son, also named Philibert, and this led to an amusing incident...

The black and white bride.

Maud arrived from Seychelles to marry Philibert at the end of 1923. She had seen a photograph beforehand and when they first met she liked him though it was not love at first sight. The date was set for February the 9th the following year.

Grandpapa Ducrotoy decided he would take the bride to the church in his car rather than a carriage and no-one dared argue with that decision. He then owned a Model T car four years old, nonetheless, it needed a coat of paint. Grandpapa thought another coat of black would be appropriate and each day the family waited for the paint job to be undertaken. The wedding date came closer and closer and the mild reminders to Grandpapa about the painting brought forth grumbles and his moustache bristled as he retorted that there was plenty of time. The wedding preparations were otherwise progressing well, the dress, cake and the family journeying from various distant regions.

Grandmama and the daughters were extremely worried about the car but they were too nervous to mention it again and risk the wrath of their father. The night before the wedding, seeing that the car was still unpainted they resigned themselves to the fact that it would remain that way. On the morning of the wedding Grandpapa fetched his brushes and started to paint inside. Still too timid to say anything the girls watched in horror as he painted inside and out, dabbing little finishing touches here and there.

In those days there was no quick drying paint as today, the old style paint took days to dry even in a hot climate. Three o'clock came and Maud donned her white wedding dress and the whole family trooped out to the car. Grandmama had had the forethought to put down white bed sheets around the still wet interior. Maud sat gingerly in the rear of the car too frightened to move or make a comment and tried to appear quite normal and relaxed. Hot tempered Grandpapa could still be enraged and terminate the wedding arrangements.

The day the pigs came home...

Grandpapa had a small farm where he raised chickens and pigs for the family consumption as was the custom in those days. One day he travelled up into the high country of Kenya and brought back to Mombasa many pigs that he had purchased. The only way to travel with them was by train and they were duly loaded into the goods wagon. After many miles and leaving the cooler countryside the temperature started to rise and the atmosphere stifling, not helped by the many delays en route. Becoming concerned for the livestock he decided to inspect his pigs and made his way down the train finally jumping into the wagon with them. He found eight pigs choking in the heat and breathing their last. His, at best, mercurial temper, was not improved when on re-joining the passenger carriage there were complaints. The seated passengers, impeccably dressed in their starched suits complained bitterly that he stank of pigs. He cared not a jot for

passengers and sat worrying until the end of the journey.

On reaching home he rushed into the house and summoned the servants and all the daughters of the family. Instructions being to light a fire and boil as much water as possible to prepare the pigs that had died on the journey. The entire household was awake until long after midnight, cooking, salting, drying the pork and preparing the offal with preservatives for sausage and other savouries. The men cut the joints after skinning the meat. The family and neighbours were well fed for months.

Some of the pigs survived and they were greatly loved by grandpapa and he was most proud of them. Anyone showing the slightest notice of these smelly animals went immediately to his heart...except on one occasion

The Greek and the German suitors...

In these days of political correctness and enlightened times it is intriguing to consider how people once spoke their minds, in particular Grandpapa...

It was following the First World War and Edma, Octavia and Paquerette, attractive girls in their twenties, were sitting in the shade of the veranda at the family home in Makupa. Two young handsome men riding along in a pousse, a hand drawn rickshaw, rode up and down the road. Finally the pousse came back down the track and around to where Grandpapa was tending his pigs. They introduced themselves and Grandpapa was flattered that the men were interested in his favourite pigs and he spent a considerable time explaining the intricacies of pig rearing. After a time they shook hands and departed and the three girls watched them depart in despair, they had hoped to catch a groom.

Grandpapa was out one day when the two men returned, one a swarthy gentleman from Greece, the other a sandy haired strapping lad from Germany, knocked at the door. The three girls were delighted to welcome the visitors, although they were unchaperoned except for the servant, and invited them in for coffee. A lively conversation followed and lots of laughter.

After some time Grandpapa came home in his carriage and entered the house. When he saw the two gentlemen he said, "Who invited you into my home?"

The girls, embarrassed, tried to explain that the men came and they invited them in, when Grandpapa swept them aside and shouted, "Get out of my house." As the men apologised he shouted again, pushed the men to the front door, and with the last war still fresh in his mind he roared, "Get out you pig of a Greek and dirty Boche!"

The crestfallen girls, Edma, Octavia and Paquerette, did not share his racist views and were bitterly disappointed that their plans to catch a husband or two had gone astray...

Grandpapa died at the age of fifty nine years. He had received a blow to his chest some months previously and it was thought that this caused the pneumonia which killed him.

MAKUPA, MOMBASA 1931.

The house surrounded by coconut palms at Makupa suburb of Mombasa built by Philibert Ducrotoy on his arrival from Seychelles. DeWet Savy is standing in the doorway. The Ducrotoy children are: Rita with baby René in pushchair; Albert on the left and Willie on the right aged two years. (1931)

THE DEATHBED SCENE 1924

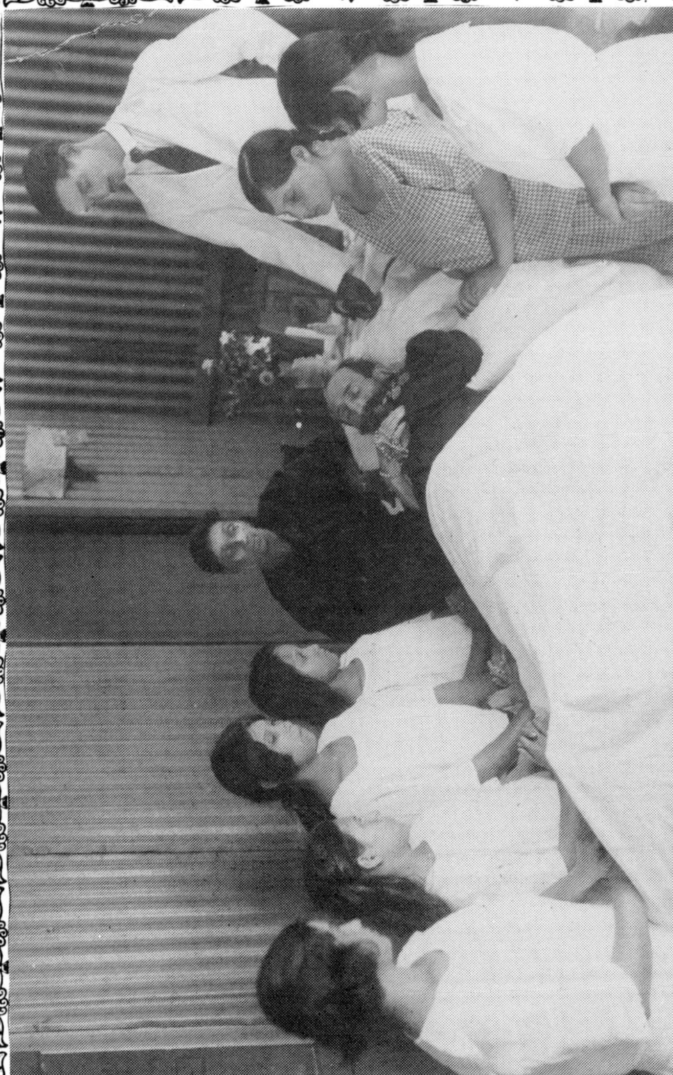

Philibert Olivier Julefy Ducrotoy. The taking of photographs at someone's death was a custom at this time but this photograph was forwarded to Philibert's relatives in France.

CHAPTER 3

DEATH IN MOMBASA.

Philibert Olivier Julefy Ducrotoy deceased.
This photograph, thought to be the only one in existence, taken for the family in France and sent by Philibert's wife to them when he died. In her own neat handwriting on the reverse was written, "Voici notre cher defunct - sons fils nois filles - et mois aussi - Souvenir sanglant. (This is our dear deceased -his son and his daughters - and me also -Sad remembrance.)

Left to right: Octavia Ducrotoy, Leonie Frost (his grand-child) her mother Bella does not appear in the photograph, Edma Ducrotoy, Julia nee Edmond, the widow, the deceased Philibert Olivier Julefy Ducrotoy 1865-1924. Philibert Ducrotoy the only remaining son, the other son Joseph having died many years before, Maud Ducrotoy, nee Savy in the check frock and expecting her first child Rita; lastly, Paquerette Ducrotoy.

In close detail the photo shows the coverlet embroidered with flowers, the pillows are white lace, the deceased is holding a rosary and wearing his First World War medals. These are :
1914 - 1915 Star in bronze.
1914 - 1920 British War in silver.
Victory Medal in bronze.

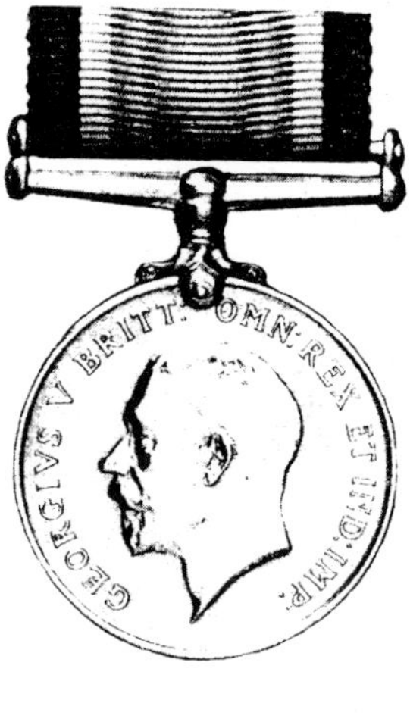

The Victory Medal has inscribed on the reverse, "The Great War for Civilisation". Philibert senior was forty nine and young Philibert nearly twenty when they joined the British Army during wartime. They both received the same presentation medals. The 1914-15 Star was awarded to all those who had served in a theatre of War between 5 August 1914 and 31st December 1915. The figures 1914-15 are inscribed on the centre scroll. The ribbon is identical to that of the 1914 Star. The British Medal awarded to them for serving outside of Britain between 5 August 1914 and 11 November 1918. The medal is made of silver and the ribbon has an orange watered centre with white stripes of white and black at each side with borders of royal blue.

THE WORLD WAR 1 MEDALS AWARDED TO PHILIBERT OLIVIER JULEFY DUCROTOY
(senior and junior)

Victory Medal, 1914-1915 Star and 1914-1920 British War Medal.

Name	Sqdn.	From	To	Remarks
Deakin, R. H. ...	E.	28/10/14		
De Cerjat, F. ...	B.	10/ 8/14		Killed 3/11/14.
De Jager, F. J. ...	D.	7/ 8/14	5/ 1/15	
De Jager, P. J. L. C. ...	D.	7/ 8/14	21/ 1/15	To E.A.T.C.
De Waal, E. H. ...	D.	7/ 8/14		L/Cpl. 26/9/14.
Dedonkele, R. R. ...	E.	28/11/15		Killed 19/1/16.
Dennis, J. G. ...	B.	26/10/14		To Scouts; wounded; taken prisoner 21/9/15.
Dent, C. ...	E.	8/ 8/14	26/ 9/14	To K.A.R.
Dent, R. G. ...	B.	8/ 8/14	15/ 9/14	To K.A.R.
Dickens, W. H. ...	C.	12/ 2/15		To E.A.S.C.
Dobbie, H. ...	B.	19/10/15		To E.A.T.C.
Dobbin, H. W. L. ...	B.	27/ 1/15	27/ 4/15	
Doble, F. J. ...	E.	8/ 8/14	12/12/14	
Doherty, H. S. ...	C.	6/ 3/15		
Donnelly, S. ...	B.	8/ 8/14	29/ 8/14	
Douglas, A. ...	C.		9/11/14	To E.A.T.C.
Douglas, W. ...	B.	9/ 9/15		To E.A.S.C.
Douglass, F. W. ...	B.	26/10/14		Lieut. 15/11/14; Capt. 1/2/15.
Dovey, G. C. ...	C.	6/ 8/14		To Scouts.
Drake, T. H. ...	B.	8/ 8/14		Killed 3/11/14.
Driscoll, B. O. ...		31/ 8/14		
Drought, J. J. ...	E.	15/ 1/15		Lieut.; from Ross's Scouts; Capt. 1/9/15.
Drummond, F. G. ...	B.	8/ 8/14		Killed 3/11/14.
Drury, N. C. ...	B.	15/11/15		
Dry, T. A. ...	D.	27/ 8/14	21/ 1/15	To E.A.T.C.
Dry, G. C. ...	D.	27/ 8/14	21/ 1/15	To E.A.T.C.
Ducrotoy, P. ...	E.	11/ 2/15	6/12/15	Wounded 13/5/15; to E.A.M.S.
Duffy, J. F. ...	B.	25/ 8/14	24/10/14	
Duirs, A. B. ...	B.	5/ 8/14	1/ 2/15	Lieut. 8/8/14; Capt. 30/10/14.
Duirs, E. G. ...	E.	8/ 8/14		To Scouts.
Dunman, V. W. ...	B.	8/ 8/14		Cpl. 10/8/14; Sergt. 21/11/14; Lieut. 6/12/15; to K.A.R.
Dunn, A. H. ...	C.	7/ 8/14		
Du Plessis, N. J. ...	A.	24/ 9/15		
Du Preez, I. J. ...	D.	13/ 8/14	24/12/14	To E.A.T.C.
Du Toit, J. J. ...	D.	7/ 8/14	21/ 1/15	To E.A.T.C.

MINISTRY OF DEFENCE
Army Medal Office Government Office Buildings Droitwich Worcestershire WR9 8AU

Telephone Droitwich 772323 } ext 44
Army Network Droitwich Mil }

Mrs C Ducrotoy
Primrose Cottage
Kingsdown Road
St Margarets-at-Cliffe
DOVER
Kent CT15 6AZ

Your reference

Our reference
D/Medals/1/1/6Y
Date
21 May 1984

Dear Sir

I refer to your letter dated 21 January 1984 and your reply to our questionnaire dated 29 February 1984.

In the absence of an Army number it is not possible to make a positive identification but it is thought that one of the following, recorded on the 1914-18 War Medal Rolls, could refer to the subject of your enquiry:-

No. 557 later No. 6041 Pte Philibert DUCROTOY - East African Mounted Rifles and East African Medical Service

Awarded - 1914-15 Star)
British War Medal) - issued 22 June 1922
Victory Medal)

Entered Theatre of War - Africa - on 11 February 1915

No. 1752 Pte Philibert DUCROTOY (Jnr) - East African Transport Corps

Awarded - 1914-15 Star)
British War Medal) - issued 22 June 1922
Victory Medal)

Entered Theatre of War - Africa - on 22 November 1915

I have enclosed an explanatory list of medals instituted to recognize 1914-18 War service, which may be of interest to you.

I am sorry I cannot be more helpful as the service records of both these men are no longer available, having presumably been destroyed by enemy action in 1940.

Yours faithfully

R. Bowles

Officer in Charge
Army Medal Office

The Victory Medal awarded to them for serving in the theatre of War (outside of Britain) between 5 August 1914 and 11 November 1918. The medal is made of bronze and has on the reverse the inscription, as mentioned above, "The Great War for Civilisation". The ribbon had a red centre with green and violet on either side shaded to form the colours of the two rainbows. Philibert senior had died before the commencement of World War 11 and Philibert junior was rejected on medical grounds. He had an enlarged spleen caused by repeated bouts of malaria.

Perhaps it was a sign of the changing times but none of Philibert's sons shared any enthusiasm for following in their father or grandfather's footsteps. When the call came Philibert went along, as the law required, to register his sons for future service should the need arise. At the registration desk he explained to the Medical Officer that his eldest son was retarded and could not be on the register. The doctor agreed and discharged him from duty crossing his name off with a flourish. When Philibert related the tale later that evening Willie and René both said, hoping for de-registration, "Why didn't you tell him we were mentally retarded too?"

Livid, Philibert replied, "Do you expect me to stand before the entire town and admit I'm the father of *three* deranged sons..."

The children of Philibert Ducrotoy of France and Julia Edmond of Seychelles.

DUCROTOY OF SEYCHELLES 1880

Irma Virginie Thuillier, Vignacourt, France = Joseph Ducrotoy, France

Jacques Edmond, Seychelles = Marie Henriette Phillipine Lebon, Seychelles

Philibert Olivier Julefy Ducrotoy 1866 - 1924 = Marie Joseph Julia Edmond 1873 - 1950

Bella	Henriette	Philibert Olivier Julphy Ducrotoy	Jeanne Marie Adele	Edma	Joseph	Maria Emily Paquerette	Octavia	Julia
1892-1965	1894-1958	1892 - 1924	1900-1984	1901-1984	1903-1920	1906-1992	1907-1989	1910-1952

The above shows the nine children of Philibert and Julia Edmond. Their history will be continued in chapters 10-13 but for the moment we stay with the Edmond family. Julia, as she was known, the only daughter of Jacques Edmond and Marie Henriette Phillipine Lebon were both born in Seychelles. Her mother being the daughter of Remy Lebon and Adēle Barbe.

CHAPTER 4

Julia Ducrotoy in centre (nee Edmond, daughter of Jacques Edmond) with grandchildren, Willie, Albert and Rita Ducrotoy. Julia's daughters Paquerette and Julia at rear.

THE EDMOND FAMILY OF SEYCHELLES.

The family have always understood that the Edmond family came from Mauritius. Since the Edmond family were extremely well educated for the period and held Government posts it is quite likely that they came from Mauritius or Madagascar with the French colonists in the late eighteenth century.

Grandmama Edmond was well educated in French, she knew the classics, Alexander Dumas and others, read a lot and was always writing letters. She was always correcting the children's manners and everything had to be carried out correctly in every respect.

She was born on the 21st April 1873 five o'clock in the morning at Port Victoria, Mahé. The only daughter of Jacques Edmond a dispensary clerk and Marie Henriette Phillipine Lebon a seamstress who married on 10th April 1855. She was named Marie Josephe Julia Edmond but known in later life by the name Julia.

Julia had three brothers who were also well educated. Their occupations were Registrar, accountant, and sea captain. One brother was born on 30th May 1863 at Port Victoria and named Charles Elliot Augustin Edmond. Another was Fernard Edmond. Rene recalls one of the Edmond brothers visiting Mombasa in 1935 on his way home from Dar Es Salaam to Seychelles.

Delusions of grandeur...

Unfortunately one of the brothers suffered from the mental condition of 'delusions of grandeur' and ended his life in the Mahe hospital. He thought he was a Count of noble birth and refused to eat his food unless a 'page' was standing beside him to serve. The family paid for all this extra attention. In the early days, in spite of his condition and because of his education, he was given the task of keeping the asylum accounts. When his sister, grandmama visited him in the hospital on a brief visit to Seychelles she found him in his office with a stern face. When he saw young Paquerette, Octavia and Julia in their white dresses he asked, "Who are these white herons?"

When grandmama explained that they were her daughters he shouted, "It is all lies. They are really spies sent from the Court of Marie-Antoinette."

He asked how the two Philiberts were and grandmama explained that they were away fighting in the Army, it was then 1915, and that one was a Corporal and the other a Private. At that brother Edmond thumped his fist on the desk and roared, "Why do you always put our family down? Why, only this morning I read in the newspaper that Bertbert (Philibert) was a General and the other a Colonel."

It was sad that brother Edmond refused to see his wife and children and did not recognise them.

There are references to the Edmond family in the Seychelles Archive at the British Library, London. The Blue Book Returns of 1881, Civil Establishments, mentions Jacques Edmond as being a Tide Waiter in April 1876, appointed by the Governor, with an annual salary of 1000 rupees but no housing allowance. This is very likely one of grandmama's brothers who would have been around twenty years of age at that date, though it may have been their father. Jacques Edmond was described as a dispensary clerk ten years prior to this and on the marriage of his daughter in 1881 as a proprietor of Mahé.

With that level of salary it demonstrates that the Edmond family were in a position to educate their family. As an example of prices of the time, hospital fees were 3 rupees for a first class patient, a simple grave with inscription was 4 rupees, tuition at school 2 rupees a month. In 1906 there were eight schools in Seychelles and the majority in Mahé.

The first record found so far of Jacques Edmond's work appointment was on 31st August 1858. He was born in either 1832 or 1833 so he would then have been twenty six years of age.

Seychelles was administered from Mauritius until 1903 and there is a possibility that he was transferred from there. Grandmama always mentioned that both her parents came from Mauritius, that her father was dark with straight black hair and her mother was white.

On the marriage certificate of Jacques Edmond and Henriette Lebon, Jacque's mother could not sign the register as witness, and his father, also called Jacques was not present at the marriage. He may have been deceased by this year. The certificate confirms that both Jacques and Henriette were born in Seychelles between 1832 and 1836. The parents of Marie Henriette Phillipine Lebon were Remy Lebon and Adele Barbe. If they were born in Seychelles then that would have been at the beginning of the 1800's.

Family Tree showing Edmond, Joseph, Lebon and Barbe families.

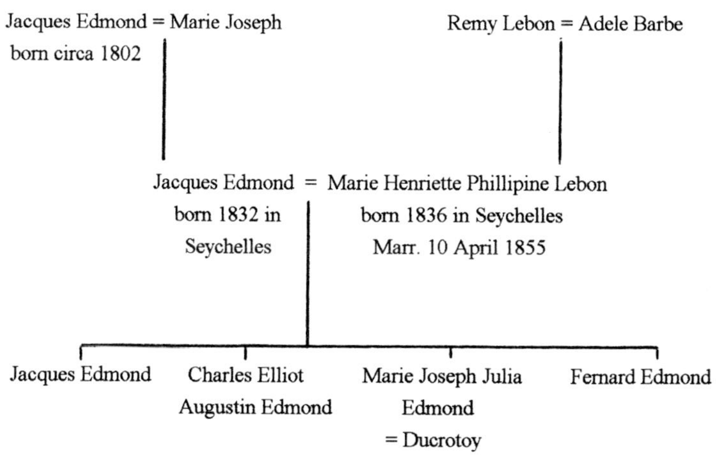

CHAPTER 5

ONE QUIET NIGHT.

Grandmama Julia Edmond related the following story to her children.

She had attended the sad funeral of a young black child of around four years old one afternoon, the child of a family living close by. All funerals are carried out the same day in the tropics due to the extreme heat and this one was no exception. Her husband Philibert Ducrotoy did not accompany her to the funeral but instead went to town and visited his drinking pals. He came home late evening and fell into bed rather worse for the drink...

As Julia was about to prepare for bed with her daughter beside her she heard a commotion around the walls of the house. There were loud whisperings such as, "There he is, get him, there he goes..." and so on. Julia opened her wooden shutters and across the clearing in the moonlight she was astounded to recognise the very same child that had been buried that day. She saw the child clearly in the white nightshirt in which he had been buried. He was running and hiding in the shadows and briefly running into the light then back into the shadow of the house as he evaded capture. Julia's daughter Bella also recognised the child and expressed surprise. The child ran off with the men in pursuit.

Julia tried to awaken her inebriated husband and even sticking a hat pin into him. She was in a mixture of fear for herself and concern for the child still in his white shroud. Philibert slept on... In the morning, after a wakeful night she recounted the strange events of the night to her husband. Philibert said, "Why didn't you wake me? I could have chased the men."

When she related the fearful story to her parents she was told that in no circumstances should she tell the authorities because if word came out her child might then be in danger. Her parents had lived a long time on these islands and it was then known that witchdoctors had a medicine that could throw a child into a coma. It paralysed them so it appeared that death had taken place. After the Christian burial attended by all the family the witchdoctor and his helpers would dig up the body and revive them followed by a ritual killing. It was thought dogs and children were killed to make gris-gris candles from the body grease.

Remains found...

The following day word was sent around that some intestines had been found on a rock near the seashore. The authorities were called and when nearby villagers were questioned they told officers that a pig had been killed there. Thus satisfied, the remains were removed and buried. The villagers however were not convinced but kept quiet. Everyone knew that when a pig is killed every part was saved for cooking including the intestines...

These chilling stories of the last century abound so it is intriguing to have an eye witness account from grandmama and her daughter.

In addition to these alarming stories there are comical fables handed down to the children such as the eel that could converse...

The talkative eel.

This tale comes from the island of Mahe and was told to grandmama Julia by her parents who claimed to know the villagers in the story. She enjoyed relating this legend to her grandchildren who, to her dismay, collapsed into incredulous laughter every time they heard it...

A young healthy family without children lived above the hills on a plantation near Victoria. The man of the house became ill with severe back trouble and lay in his bed for two years. Since he was the wage earner the couple went from being poor to abject poverty despite meagre help from their families. One by one each article of furniture was sold to eke out a living until at last even the bed had to be sold and the patient was left lying on the mattress.

One extremely hot afternoon as they both rested on the mattress, the door was ajar to allow the cool air to flow, the wife noticed a movement by the door. An eel crawled to the door and up and over the ledge and lay watching them. The woman jokingly said to the eel, "Yes, now you have made my husband ill you have come to see how we are prospering..." To her astonishment the eel replied, "It is not me who makes your husband ill but the man who comes to the house every morning. He is responsible." The eel then slithered away...

There was a man, a friend, who called each day for coffee on his way to work so the wife waited for him to arrive the next day. The time came as and he came up the path to enter the house he seemed uneasy and would not look the woman in the eye, nonetheless, he sat at the table. The woman found suddenly she had power over him and as she stared the visitor became more and more agitated and wanted to leave. The woman, after deep thought, suddenly said, "I now know you are responsible for making my husband ill." The friend tried to deny it saying he had been falsely accused and then left. The very next morning as the sun was rising the friend called at the door but refused to enter the hut. He quietly handed the wife some herbs and instructed her to make a herbal tea and give the drink to her husband. From that day her husband rapidly improved and by the end of the week he had returned to work...

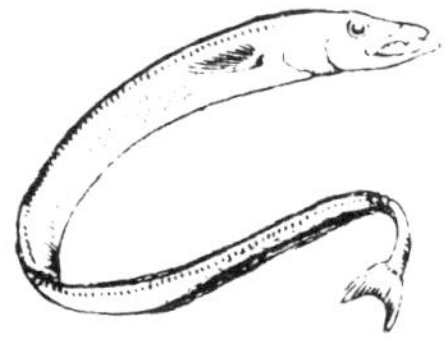

It is difficult to imagine these as bedtime stories but they kept the grandchildren quiet for years. This story passed on to grandmama from her grandparents adds a little weight to the theory that some of her ancestors came from Mauritius.

Tales from Mauritius.

A young man went to a dance and met an extremely attractive girl and he insisted on accompanying her home. She declined and said mysteriously, "Where I live you cannot go..." The man suspected she was embarrassed about the area where she lived, nevertheless, he persisted. She repeated that it was not possible for him to accompany her but on his continued insistence she acquiesced.

They both walked many miles without talking and the eventually came to the cemetery gates. She then turned to the young man and said, "This is where I live." She then walked through the cemetery gates and disappeared.

The man, completely shocked, ran as fast as his legs would carry him not stopping until he reached his family home. He collapsed when he reached the door and after relating the strange story to his parents he took to his bed. The parents called a priest but from that time on the man would not eat and he died several weeks after the incident.

Death of Grandmama Edmond-Ducrotoy.

Marie Josephe Julia Ducrotoy, nee Edmond, died at Molo, Kenya around the year 1948 she was aged seventy five years.

Grandmama had been living with son Philibert and Maud in Miwani but she suddenly expressed the desire to stay with her youngest daughter Julia. Rene, Willie, Hilda and Edmond, DeWet and Julia were then residing and working in Molo. When Hilda and Edmond (Hoareau) were married the two bachelors, Rene and Willie, moved into another house and gave theirs to them. When DeWet arrived in Molo the genial bachelors moved out once again and into a hut.

The family tried to persuade grandmama not to come to Molo saying the house was too small but principally that the altitude was so great. She had a heart condition and asthma so in view of her medical condition it would be foolhardy in the extreme. Grandmama promptly arrived...

The altitude at Molo was above 8,000 feet and for someone used to a tropical clime it was extremely cold. In the mornings the grass was covered with frost and any buckets of water left outside were frozen solid by morning. All this some thirty miles from the equator.

It is now recognised that anyone with cardiac problems should not travel anywhere over 6,000 feet in altitude. She lived just two weeks...

During the afternoon of the day she died Nadege and Neville, Julia's young children, crept into her room where grandmama lay. There were drinks and biscuits on the bedside table and they quietly reached for the plate. Grandmama opened her eyes and shouted, "I am not dead yet. You are stealing my biscuits!"

On the evening she died she begged Rene and Willie to take her back to Miwani. There was no hospital only an outpost with no oxygen available. They called the doctor and Dr Maclean arrived at eight in the evening to administer medication. He advised against any move from the town as to travel out of Molo the transport would have to climb to 10,000 feet before descending. She implored them to take her back home but even considering the medical advice there were the practical difficulties. Eighty miles of bad roads to negotiate in bad weather with a safari van. Grandmama died three hours later...

There was no Roman Catholic church at Molo and the nearest place to take the body was Nakuru. It was a long dusty ride followed by a frustrating time with official papers to sign and a sad time informing all the relatives then waiting for the family to arrive. Her son, Philibert, came alone by train and grandmama was buried in Nakuru Catholic Cemetery. The year was 1948.

Grandmama's daughter-in-law Maud Savy, the wife of her son Philibert, was descended from Charles Savy of the island of Saint Anne. It is to this family we now return...

CHAPTER 6

CHARLES SAVY OF SEYCHELLES

"It was generally known that we were directly descended from the notorious slave trader in Seychelles, Charles Savy. My family told me that he went to Africa to collect slaves to work his land on Mahe. This was a regular trip since they soon died of disease or neglect and needed replacing. Many countries carried out the slave trade in those days but when the British abolished slavery he continued to ply his trade regardless. He was well known and the British frigates sailed the coast trying to reduce the problem. When Charles Savy or his head man spotted the frigates in the distance he dropped the slaves over the side of the ship to their deaths rather than be caught with them aboard.

My grandparents told me that their ancestor, Charles Savy, was greatly feared and gave a lot of trouble to the authorities of the time.

One day he removed the roof of his house to renew the thatch using his slaves as workmen and helpers. The job was starting to take longer than anticipated and it soon became evident that a storm was on its way. Charles Savy became more and more furious and shouted at his workers and as his anger increased they became even slower and more and more confused as to his wishes. His fury at the thought of his furniture being ruined in the rain made him rush to the house and return with a shotgun. He stood guard over the slaves and vowed he would shoot every one of them if the roof was not replaced before the storm broke..."

Maude Savy Ducrotoy 1973

The following letter was received from the previous Director of the National Archives and Museum, Mr H.J.McGaw, dated 15th December 1982 regarding Charles Savy.

"Charles Savy continued taking slaves to Seychelles after the abolition of slavery as there appears to have been grievances among the owners of slaves to this effect and they secretly continued the illicit trade despite tough measures of the authority. However, since Mauritius was then the administrative headquarters I am pretty sure that the Mauritius Archives could provide you with the necessary information to establish a true fact as to whether Charles Savy was involved in shipping activities in the Indian Ocean.

Charles Dorothe Savy who came to the Seychelles from what was then known as "Ile de Bourbon" and now known as Reunion and Miss Marie Joseph Jorre de St Jorre* who came to Seychelles from Mauritius. These must have been one of the earliest settlers in Seychelles as it is noted that their eldest son was born in 1794." **

Note from author: * Amongst my mountainous heap of papers there is a copy of a land dispute relating to this Jorre de St Jorre family lodged in the Courts circa 1774. Athalie D.

** The Savy - Jorre de St Jorre family tree is within Chapter 15.

The following family tree shows the author's mother-in-law, Maud Savy and her ancestors. Maud was born in 1906 and married Philibert Ducrotoy (the younger). The chart shows her great grandfather Victor Francois Savy who married Bidale Lalanne and her grandfather Vildoric Victorin Savy who married Marie Laurestine Aimard Morel.

THE SAVY FAMILY OF ST ANNE.

Adelaide = Lalanne
born 1786-88

Francois Emmanuel | **Victor Francois SAVY** = **Bidale Lalanne 1808-1885** | **Jean 1804** | **Jeanne 1806**

Louis Victor 1836 | **Vildoric Victorin SAVY 1843** | **Jeanne Marie Louise 1838** | **Alphonsine 1827** | **Medoric 1841** | **Virginie Victorine 1846**

Victorin Livdoric Savy 1871 | **Marie Elisca SAVY 1869** - - **HenriVerlaque 1864**

Celio 1895 | **Henri 1897** | **Maurice 1898** | **Maud 1906** | **Mauricia 1900** | **DeWitt 1902** | **Filis 1905** | **Carl 1908** | **Alienne 1909**

Victor Francois Savy was born on 11th July 1798 and his wife Bidale Lalanne born at Mahé on 2nd January 1808, the daughter of Adelaide and Lalanne who were unmarried. Victor and Bidale married at Mahé on the 16th December 1858, thirty one years after their first child was born.A declaration was made on the marriage certificate which legitimised the children of the marriage. It is possible that it was suggested by the priest when one of their children applied for a marriage licence.

VERLAQUE : DAVIS

Henriette Marcienne Verlaque = John 'Bull' Davis
(unmarried)

Marie Elisca Savy + = Henri Verlaque | Francois 1866 | Dartagnon* Verlaque | James (Mimi)

Dartagnon* Verlaque: Victorine | James | Dartagnon*

*The mother of Dartagnon (either the father or the son died at his birth in Seychelles)
The above tree shows Maud's true parentage known to the family though not officially documented. Her father was Henri Verlaque and his father John Davis, a Welshman.

***

SAVY - MOREL - POILLEVERT - VERLAQUE

Herminie Poillevert = 1st husband: Furey Morel
2nd husband: Jean Baptiste Verlaque

Vildoric Victorin Savy = Marie Laurestine Aimard Morel
1843 1842

The above tree shows Maud's maternal grandparents and great-grandparents.Furey Morel died and her great-grandmother married secondly Jean Baptiste Verlaque.

The bride is Alice Morel whose mother was Victorine Verlaque, Maud's cousin on her father's side of the family.

Seychelles Marriage of Victor Francois Savy and Bidale in 1858.

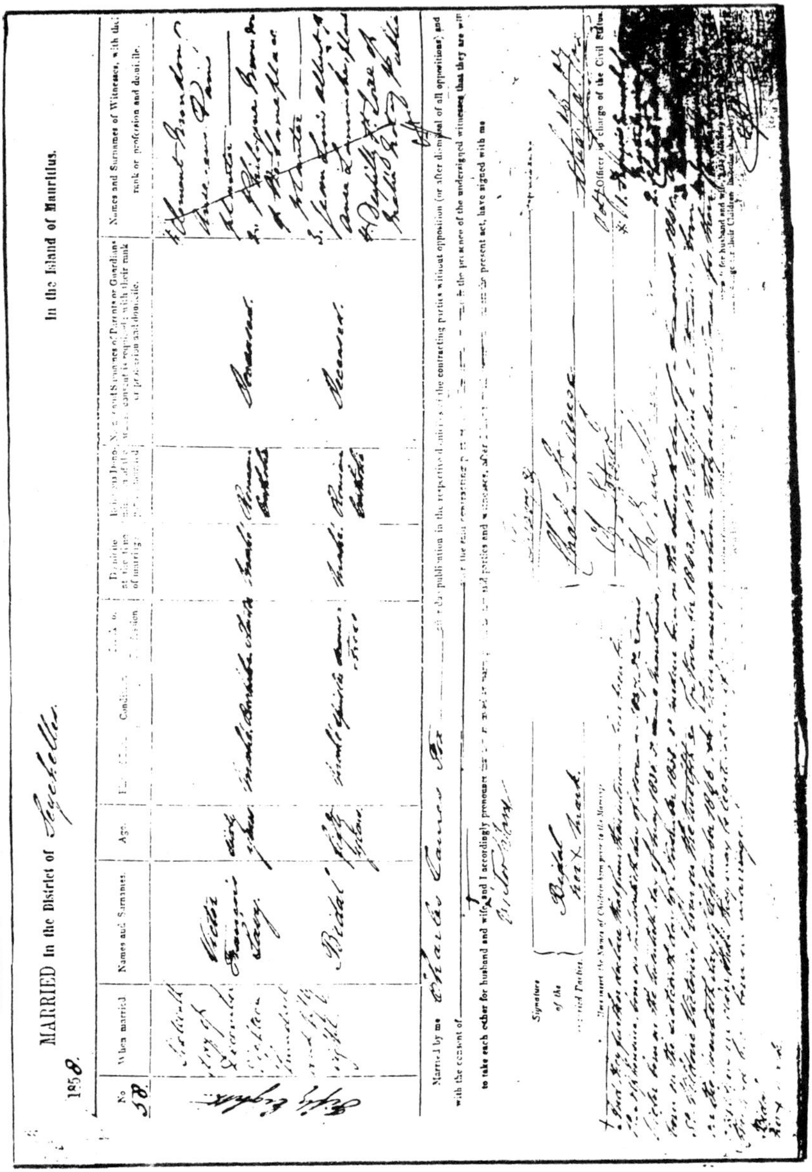
MARRIAGE IN THE SEYCHELLE ISLANDS IN 1858. VICTOR FRANCOIS SAVY and Bidale Lalane.

1858. MARRIED in the District of Seychelles In the Island of Mauritius.

No.	When married	Names and Surnames	Age	Condition	Rank or Profession	Residence at the time of marriage	Religious Denomination	Names and Surnames of Parents or Guardians whose consent is required, with their rank or profession and domicile	Names and Surnames of Witnesses, with their rank or profession and domicile
58	[illegible]	Victor François Savy	[illegible]	[illegible]	[illegible]	[illegible]	Roman Catholic	Deceased	[illegible]
		Bidal	[illegible]	[illegible]	[illegible]	[illegible]	Roman Catholic	Deceased	[illegible]

Married by me Charles Camus [illegible] after publication in the respective domiciles of the contracting parties without opposition (or after dismissal of all oppositions) and with the consent of [illegible] to take each other for husband and wife, and I accordingly pronounce them to be husband and wife in the presence of the undersigned witnesses that they are witnesses to the present act, have signed with me

Signature of the married Parties: Victor Savy; Bidal her X mark

Officer in charge of the Civil Status

16th December 1858 in Mahé, Seychelles.

The following transcript provides the details more clearly...

Marriage of Victor Francois Savy and Bidal 16 December 1858 in Seychelles.

It is fortunate for us that they chose not to marry until later in life for it has provided a wealth of information on one certificate which needs some explanation.

Victor Francois Savy, a bachelor, and planter of Mahé married Bidale Lalanne on 16th December 1858 at Mahé Roman Catholic Church. Although she is named just 'Bidal' on the marriage certificate we know from previous documents that she was the daughter of Adelaide and Lalanne of Seychelles. The bride and groom were aged sixty years and fifty years old. How it came about we can only speculate at this stage but they had lived as man and wife and produced six children.

Additions to the marriage certificate were as follows:

'And further declare that from their intercourse have been born:

Alphonsine born on 20 November 1827.

Louis Victor born 20 May 1836.

Jeanne Marie Louise born 16 November 1838.

Mèdoric born 11 August 1841.

Vildoric Victorin born 12 November 1843.

Virginie Victorine born 20 September 1846

....born before their marriage whom they acknowledge for their children in order that they be legitimated as if they had been born in marriage.

Bidal signs with a X being unable to sign... Bidal was a seamstress, and at her death it is stated she was a proprietoress Mahe. By the time of the marriage both of their parent's had died. Witnesses to the ceremony were Francois Emanuel Savy a Master Mariner of Mahe, and Charles St Ange of Mahe,

- - - - . Charles - - - of Mahe. Joseph Guillaumain, Master carpenter of Mahe.

René mentions that illegitimate children were baptised and married during the week and not at the usual time in front of the congregation . The couple may have decided to legitimise the children when one of the children were planning marriage. More investigation in the Seychelles Archives is required.

Children of Marie and Elisca Savy and Henri Verlaque of Seychelles...

Bidale Savy died at 2pm on 31st March 1885.
The children of Bidale and Victor Savy were as follows:

Vildoric Victorin Savy born 12th November 1843
Jeanne Marie Louise Savy born 16th November 1838.
Alphonsine Savy born 20th November 1827
Louis Victor Savy born 20th May 1836.
Medoric Savy born 11th August 1841
Virginie Victorine Savy born 20th |September 1846

***

Vildoric Victorin Savy was a carpenter and married Marie Laurestine Morel who was born in 1842, the daughter of Herminie Poillevert and Furey Morel. At the time of her marriage her mother had married her second husband, Jean Baptiste Verlaque.

There were other children of Vildoric and Marie's marriage but so far only the following have been found in the archive
Victorin Livdoric Savy born 21st March 1871.
Marie Elisca Savy born 5th June 1869.

***

Children born to Marie Elisca Savy and Henri Verlaque (unmarried) were;
Celio Cebert Savy born November 1895.
Henri Furey Savy born 1st March 1897.
Maurice Francois Savy born 24th December 1898
Mauricia Therese Savy born 1900.
De Witt James Savy born 1902.
Louis Filis Savy born 23rd July.
Carl Ronald Savy born 1908.
Alienne Regina Savy born 1909 and died 1951.
Maud Vivienne Savy born 3rd March 1906 and died 1974

***

Maud Vivienne Savy born 3rd March 1906 married Philibert Olivier Julefy Ducrotoy on 9th February 1924 at Mombasa.

Children of the marriage:
Rita Julia Ducrotoy born 2 January 1925, Mombasa.
Albert Reginald Ducrotoy born 26th May 1928, born Mahe in Seychelles
Philibert Olivier Julphy Willie Ducrotoy born 29 June 1929.
Maxsim René Ducrotoy born 3rd November 1930, Mombasa.
Hilda Elisca Ducrotoy born 5th August 1932, Mombasa.
Mary Mavis Muriel Ducrotoy born 25th March 1933.
Iris Patricia Ducrotoy born 29th November 1935
Winnie Blanche Ducrotoy born 19th September 1937,died 1939.
Nancy Agnes Mary Ducrotoy born 7th October 1938.
Stella Maureen Dorothy Ducrotoy born 20th May 1942.

That this book is written at all is due to Maud. She was then living in Brisbane and Rene and I on the remote island in the South Pacific, New Hebrides, which is situated midway between the Solomon Islands and Fiji. Rene was on contract to a local firm so was unable to travel so I flew to visit Maud with her latest grandchild Chantal who was the eight months old. We sat for hours talking of her early life on Saint Anne in Seychelles. Her hardships of early life, sad and then amusing moments of life inspired me to make copious notes and these pathetic scraps of paper have been carried all about the world in our travels. This is her story...

Maud.

Maud Vivienne Savy was born on 3rd March 1906 at one o'clock in the afternoon on the island of Saint Anne, Seychelles. The midwife was Marie Joseph Flore nee Julie, of Cascade, Mahé, who registered the birth nine days later. Maud's mother only, Marie Elisca Savy, was named on the birth certificate. Her father was Henri Verlaque which needs some explanation.

Henri Verlaque and Elisca (as she was always known) were friends growing up but her parents Vildoric Victorin and Marie Savy disapproved of the match. Elisca was broken-hearted but obeyed her parents who sent her to a relative on Saint Anne Henri departed and became a coastal Captain and he later married. The marriage lasted a year and there were no children of the marriage.

Henri went to Saint Anne Island to search for Elisca and although her parents still protested she eloped with Henri. The Roman Catholic Church did not allow divorce for Henri so they lived as man and wife and nine children were born to them.

Elisca Savy was always known as Mrs Verlaque.

An insight into his character is illustrated by an early memory. Maud remembers her father waking them up very early at dawn to leave the house and go to the beach to watch the sun rise over the water which was a breathtaking sight which she never forgot.

As children Maud and her brothers and sisters would play amongst the cacao plantation close to their house. The lady who owned the trees told them they could eat the fruit providing they return the seeds for drying. The seeds were laid out on a stone slab to dry before being ground into cocoa.

There were many Verlaque cousins who visited and Maud knew her paternal grandparents. There were twins in the Verlaque family, this may have been amongst her father's siblings.

Blonde curls...

Maude's grandmother, Henriette Marcienne, was known as 'Mrs Bull'. her husband being John 'Bull' Davis but she was from the Verlaque family of Saint Anne. It is said her family owned most of Saint Anne, the original settlement of Seychelles when the main population moved to the capital of Victoria.

Henriette Marcienne Verlaque was a blonde lady and suffered from a cleft palate which affected her speech. She could not write and she had a unique way of ordering her supplies which would be booked to her account. The servant would be told what supplies to fetch from the store and being unable to sign a letter of authority she snipped off a small blonde curl and placed it within an envelope. The store manager would open the envelope and since she was the only one around with such blonde hair he would accept the order and envelope as her personal guarantee. After some years the servant had the idea of collecting her hair from her brush each day and consequently managed to run up quite an account before the fraud was discovered...

Watching the workers.

One of Maud's relatives had a glass eye. Crystavel Verlaque yearned to go to town and he resented having to stay and watch the field workers yet whenever he left there was little progress with the job in hand.

One day he took out his glass eye and with a flourish placed it on a field post. He turned to his workers who watched in awe and said, "Now I am watching you so you are to work very hard."

Off he went and met his friends and late afternoon returned to the field to dismiss his workers. He picked up his eye and popped it in the socket and surveyed all before him.

"Yes, yes, I can see you have worked hard, you can go now."

Paralysed.

When Maud was around five years old bad times fell on the family and led to many years of extreme hardship and this is how it all came about.

It was around 1911 and the evening oppressively hot so to keep cool Henri decided to walk to a cool nearby stream. There was a coconut tree fallen over the stream and he sat on the trunk and dangled his legs in the cool running water. Shortly after he had difficulty walking and extreme pain and he was unable to walk. He was paralysed for five years and remained in bed during that time. He had feeling in his legs though and it does sound like a prolapsed intervertebral disc. Over the years, and one by one, the furniture and valuables were sold to pay doctor's bills. The family lived in profound poverty and Maud remembers the townsfolk leaving food parcels to help them. Elisca had a wealthy aunt, who was also her godmother, but she had quarrelled with her some years previously.

Squeaky shoes.

After five years Henri improved enough to undertake a little work in the home. He taught himself to make shoes and he became proficient in that. Now, popular at the time were shoes that squeaked. It was a sign of wealth that someone could afford new shoes and people would proudly walk into church squeaking away merrily and people would turn their heads to see who it was... Henri capitalised on this and found that by placing several large, thick leaves between the upper and lower sole it produced an excellent squeak. These shoes were in demand and this sustained the family for a time.

Elisca became embittered when as soon as Henri had returned to full health he left her and taking eldest son Ronald with him he set up home elsewhere. They had not reconciled when he died in 1941, some twenty years later.

Fleur d'orange.

The aunt and godmother with whom Elisca quarrelled was the subject of several stories. The aunt was weathy and employed many people. One evening one of the workers came to her and said, "Madame, my child has fever and I need fleur d'orange."

The aunt, busy doing something, told him she did not have any... The next day she learned that the child had died in the night and she was overcome with remorse. She took the three full bottles of fleur d'orange imported from France and in anguish she broke them on the stone steps. She then told her husband, "Now, I do not have any."

Prison for a pineapple.

Elisca's aunt watched one afternoon as a man passed her garden with his pregnant wife. The wife turned back and pointed to a large juicy pineapple growing in the front garden which she obviously found irresistible. The man shook his head and they passed by. A short while after the man came back, picked the pineapple and rushed off. The police were called and the man was sent to prison for two months...

Maud sewing and singing.

Maud had been to school for a short time but when her father became ill school was stopped because the fees could not be afforded. Consequently Maud was only able to sign her name. Whenever letters had to written she dictated them to her children who would write and send them. Her signature was very clear and neat and it is a great pity that she could not express herself on paper.

Maud was a wonderful seamstress and could make any clothes and never use a pattern. She could just look at an item and copy it. She spent most of her spare time sewing for her family, grandchildren and everyone.

Maud had, by all accounts, a beautiful singing voice and would sing at parties and weddings. At soirées she would rise and sing the old French songs to all assembled.

Maud and her brothers and sisters were handed down stories as children from a variety of sources and these were a blend of ghost and treasure legends. However, a search of official records have proven the reality of some of these. Maud remembers sitting around an oil lamp one evening, the light casting shadows on the wall, listening to her father recall his experiences.

Apparition on the balcony.

Maud, an excellent raconteur, also had a great sense of humour. An example of this happened during last year of her life. She had been admitted to Brisbane Hospital and the family went daily to visit her but on this particular day the general decision was to stay home.

Our children were in bed and Nancy and Philip Savy, Albert, René and Norman Jorre' sat back to watch the late film on television after an exhausting day.

We sat in the darkened room on the still, hot night with doors and windows wide to catch the breeze and watched a most eerie ghost film on television. During the interval everyone started to feel guilty about not visiting Maud. One said, "We should have made the effort she was a lot worse when we last saw her." Another replied, "If she had died we wouldn't know, not having a telephone." And so it went on.. One rationalised the situation saying, "The hospital is too far, parking difficult, the children were too tired in the heat." It was then generally agreed that the correct decision had been made. With clear consciences we settled down in the darkness with bated breath for the second half of the movie when a white vision of Maud appeared at the open window. We could clearly see her, completely motionless with the street light casting a halo of light around her silhouette. The women screamed and even the men yelled out in fright. Even Maud had a shock at the violent reaction and then started chuckling. It seems that she had made a rapid recovery that day and the medical team discharged her, subject to her family collecting her. As it happened a friend called to visit whose car Maud commandeered for her journey home. She told her friend to park along the road so we would not hear a car arrive. She walked alone, although they attempted to dissuade her, up the path quietly in her slippers and on to the veranda where she made her ghostly appearance...

CHAPTER 7

GRANDPA SAVY IN MAHE.

Maud's father late last century lived in Mahé and Saint Anne. He awakened in Mahé one morning whilst it was still dark and set off to milk a neighbour's cow. The neighbour having suddenly become ill through the night and could not attend to his duties. It was normal, especially where livestock were concerned for neighbouring farmers to help each other in this way.

Papa Henri had had a few drinks the night before but now considered himself quite sober. As he ambled past the cemetery at 5am he passed another man whom he thought was similarly going about his business. He suddenly noticed that while in most respects the man was normal his feet did not touch the ground as he walked.

Papa Henri asked the man, "What powers do you possess to enable you to walk above the road like that?"

The man glared back silently and turned and walked into the cemetery gates...

Henri Verlaque continued to his neighbour's farm and carried on with the milking. He became nervous only after he had finished his chores and there was time to reflect on the strange experience. He then felt the urgent need of someone to talk to. He called on a friend and related the story.

The friend shook his head and said, , "Oh Mr Henri, you walk too much at night. You must go to your house long before it is dark..." No-one can recall whether he listened to that advice...

The following tale was related by a friend, who it was said to be, the father of the boy involved...

The old man in the hut.

Some way out of Victoria there was a plantation and upon it there lived an old man with a rather ramshackled house though some referred to it as just a hut. He was rather elderly and certainly not able to attend to any of the heavy gardening and maintenance that was required, nevertheless, the garden was always neat and well tended.

A youth was passing along the road one day and since his father was a friend of the old man he helped him carry a large bag up the hill. The old man was slightly intoxicated and was having difficulty with the steep climb. When the hut was reached he asked the younger man to have a drink of rum with him.

It was the first time the youth had visited the hut and upon entering he was astonished to find that although it was plain the interior was immaculately clean. The old man took a glass from his cabinet and sat with his back to the door leading to another room. Down on the polished floor the large bag burst open to reveal stale bread and cakes. The men sat chatting, now and again sipping their rum, and gradually the old man became more and more drunk and began to relax. The young man slowly became aware of a babble of voices coming, not from a few people,but a great many. The old man angrily banged on the floor with his stick and shouted for quiet. This happened several times and each time the murmuring ceased but momentarily. Eventually the old man arose and went to the door with the bag which opened partially. The young man stared in disbelief as a thin ghostly hand with the sleeve of a nun's cloak grasped the bag of bread before the door was closed shut...all was then silent.

With that, the youth leapt out of his chair and ran all the way home without stopping and as came to the door he fell inside and fainted. His father shook him and said, "Where have you been that you return home in this state?" The son, after he had recovered recounted the tale and his father warned him not to visit that hut again.

The following day the father visited the old man and said, "I know you are involved in gris-gris and if you ever come near my son again I will kill you."

It was reputed that the old man practised witchcraft and encouraged the 'living dead' or ghouls to work on the property at night and this was carried out with great skill and precision. They would leave their graves at night and attend to his needs.

This well known superstition led families to bury their relatives with a bag of salt. Salt apparently having a similar effect to the well known garlic remedy in folklore...

The Laverven bush..

This legend comes from the island of Mahe... There is a bush that grows into strong sticks similar to those used for walking sticks. In witchcraft it is taboo since it is known to take away power.

During Lent one family sent their three girls to morning mass but to enable them to reach the church in time it was necessary to leave at five before it was light. As the girls reached a bend in their path they came upon a large rock. Around it they went and as they reached the other side they saw a dog with five young puppies walking in single procession. To their amazement the dog was walking upright and holding a bell as if to say mass. The children screamed and ran home to their father. Outraged, but a little nervous at the unknown, he went to the Laverven bush and cut himself a strong stick and retraced the path taken by his young daughters. Once more, as he rounded the rock, the dog came out in procession with his bell, the father lifted his stick and brought it down hard onto the dog's back.

Several days later the father, not having seen the dog's owner decided to pay him a visit. As he approached his property he noticed that the animals had not been attended to and the house taken on a neglected air. The owner of the dog, which was nowhere to be seen, was lying in bed unable to move. The father turned on the man and said, "Why did you allow your dog to frighten my children?"

The old man told him that he had only intended to play a joke on the girls. When the father examined the man's back he found he had a broken spine...

THE PRESIDENT OF DRUNKARDS.

The year was 1935 at Port Victoria on Mahé. It was a hot, humid afternoon and inside the Courthouse tempers were frayed as the Court sitting was prolonged. The prisoner, who had been found drunk on a public road the day preceeding, stood glumly in the dock awaiting his fate. He knew he was in big trouble this time as the Judge, sweating with the heat of the day, mopped his brow and read through his forty one previous convictions.

"You are a drunk", the Judge announced, stating the obvious, "You were born drunk, in fact you are the President of drunkards... you are fined twelve rupees."

An exhumed body

Many years ago, on a remote island, there was a happy couple with seven daughters and one son. The daughters, one by one, were married off and the son stayed on the plantation to work on the land.

Late one afternoon the parents watched him in the distance as he returned on the back of a donkey and they sat and waited for him. Shortly after the donkey arrived at the house but when they found he was not upon it they set off to search for him along the track. They found him lying on the ground gravely ill and before the night was over he was dead.

Several days after the funeral, gossip was put about that the son had been poisoned due to jealousy. The Medical Officer ordered the body to be exhumed and a post mortem was then undertaken. Traces of a certain substance was found in the stomach and several people were arrested. On analysis though there was insufficient poison for a correct diagnosis so reluctantly the accused were released. He was twenty eight years old...

The bountiful births

There was a remarkable woman of La Digue who had eleven children. What was unusual about that you may ask? Well, this young woman had been married but three years...She had had three sets of twins, two single births and then triplets, two boys and one girl, all healthy children...

CHAPTER 8

THE RELUCTANT DONKEY.

When Maud Ducrotoy was first married they lived on the edge of a large coconut plantation. She was sitting in the shade one hot afternoon busy with her needlework when she heard talking and the clopping of hooves approaching in the distance. Her father-in-law, Philibert Ducrotoy, had arrived with his servant and donkey to collect coconuts. Upon the donkey's back was strung two sacks to carry away the nuts. This surprised Maud since she had only ever seen coconuts taken away by cart but she watched with growing interest. Philibert came inside the house for refreshments and left the servant to load the donkey. When it was time to leave he handed the reigns to the servant and told him to hold steady. Just at that point the donkey, becoming bored, started bucking and out popped some coconuts.. Papa Ducrotoy shouted to the lad, "Hold onto the reigns." It was no use. The donkey kicked so hard the boy let go. By this time Philibert was furious and shouted orders not to frighten the donkey by chasing him. The donkey became more and more hostile and the servant, in his confusion, ran around the donkey in fear of the consequences, both from donkey and master. Whenever he came near the donkey kicked and ran even further. At each kick more and more coconuts popped out of the sacks so carefully loaded.

As Philibert became redder in the face he shouted instructions in French, then Creole, but the servant only understood Swahili. Maud stood on the verandah and laughed until she almost cried.

Maud's father-in-law still failed to see the funny side and that evening he related the tale to his wife and said, "Maud really is still a child...

THE GIANT OF SEYCHELLES.

Maud Savy, my mother-in-law, often spoke about a distant relative on her father's, the Verlaque side of the family though he died some years before her birth. She thought he was a second cousin, I have yet to prove a relationship through the Archive but she knew the family story well. Two stories were told to her as a child.

On one occasion the giant, over seven feet tall, who was known to be bad-tempered, hit out in anger and with two fingers broke a man's ribs severely. On another occasion a servant was cheeky to the giant's mother and this so enraged him that he gave the servant one slap. For his insolence the servant received a broken jaw.

Rumour went around that the giant was poisoned by the villagers who were frightened of his strength. The family, however, had a quite different view and one that is feasible. Maud was adamant that the family always stated that the doctor inadvertently gave him an overdose of medicine which killed him.

The washerwomen were too frightened to lay their washing upon the giant's grave at Bel Air as they did with the other tombs. It is still said that his ghost wanders at night in Victoria and is sometimes seen, once by a tourist. It was reported in the local newspaper many years ago that a Londoner going for an evening stroll to explore the town of Victoria found himself being followed. He quickened his pace and the person behind dropped back, at which he suddenly glanced sideways. To his astonishment and dismay and then becoming frozen with fear he looked at an enormous different view and one that is feasible. Maud was adamant that the family always stated that the doctor inadvertently gave him an overdose of medicine which killed him...

The washerwomen were too frightened to lay their washing upon the giant's grave at Bel Air cemetery as they were in the habit of doing with the other graves. It is still said that his ghost wanders at night in Victoria and has been seen, once by a tourist. (Jackson 1974). It was reported in the local press twenty years ago that a Londoner going for an evening stroll to explore the town of Victoria found himself being followed. He quickened his pace and the person behind dropped back at which he suddenly glanced sideways. To his astonishment he looked at an enormous male figure towering above him and froze with fear. The tourist screamed out and ran off as fast as his legs would go. After some distance he stopped for breath and glancing back he saw the person who had been strolling behind him also running but at even greater speed in the opposite direction.

Salted bats.

In 1850 Maud Savy's grandmama knew an old slave, a woman who dribbled from the corner of her mouth which was scarred. She had been owned by a Mrs Savage. When the slave was young she had been given the task of salting and drying hundreds of bats in the sun. These bats used for food are still considered a delicacy in Seychelles. It was the slave's job to look after and care for them, turning them in the sun but one day she forgot and it rained heavily upon them spoiling the meat. The bats were completely soaked and when Mrs savage saw this she became angry. The slave was made to eat them raw as punishment and she was forced to eat and eat until she was sick. When she stopped eating she was beaten again until she continued. The slave told Maud's grandmama that she could not face eating bats or salt fish ever again and her mouth was damaged as a result...

[Fruit of the Cacao-Tree.]

Saint Louis mudslide.

Maud remembers her family speaking of a devastating mudslide at St Louis above Victoria town in 1862 when many friends died.

Before the tragic day it rained continuously for three months, so much so that special prayers were said in church to stop the deluge. Eventually the landslide came with a roar and filled in part of the sea burying all in its path. Rescuers were hampered, not only by the driving rain and flooding but the absence of any ships in the port where the crew could offer manpower and tools to assist.

The priest of the church had a long white beard and it was that which saved him. As he became buried with mud and timber his beard came up and over thus protecting his face enabling him to breathe in the air pocket. Others weren't so lucky and seventy five people lost their lives in the disaster.

A noble family.

A member of the French nobility settled in Seychelles and married a black woman and had children. Many years later and after the French nobleman had died papers came from France to say he had inherited property and great wealth. French relatives arrived and on seeing the black grandchildren expressed doubt that they could be the descendents of their relative and returned home.

Everyone on the island knew they were the children of the nobleman and legal representatives once more visited the island. It was suggested that they should produce some evidence of the family connection, a silver buckle, a sword, silver tray or Family Bible. Owing to hard times the family had been reduced to such poverty that slowly they had sold each precious heirloom handed down to them.

The family did not inherit the family fortune...

Wartime submarine.

A German submarine visited Seychelles during the last war. The were in deep water offshore not realising perhaps that the tropical sea is so clear. It wasn't long before they were spotted from the hills by the Seychelles Defence Force. Seven men were subsequently arrested and four men escorted them to Kenya to hand them over to the authorities there. One of those four men was a relative, a Mr Berlouis.

The treasure pots

There was a very poor family eking out their existence for many years on a barren piece of land on one of the islands. They were an ordinary lot and led a simple life without coming to public attention. Not surprising then that whispering started when the man of the household went to France for a holiday. There were new clothes, horses and all manner of things that life could offer. This caused much speculation and gossip. The story was as follows...

The landowner had sent his workmen to dig holes for fencing. Soon after he set off to the site and as he arrived upon his horse a workman came to him and said, "I'm sorry master, but my spade dug too deeply and hit one of your clay pots." The boss was mystified but tried not to show it, he just nodded and sent all the workmen to dig in a different area. That night the boss returned with his son to the same spot and with the aid of a lamp started to dig, long into the night.

No-one knows exactly what was found but the family prospered from that time onwards...

Saint Anne Island coins.

One evening at dusk a ship was seen to anchor some distance from the shore of Saint Anne. Somewhere around the year 1878.

The following day Elisca Savy was working in the garden when her mother, Henriette, came running to her crying, "The dead has risen, the dead has risen...". She could not be comforted.

Henriette passed the beach each day and over the years was in the habit of placing flowers on an unmarked grave, a stone slab, as a mark of respect for what she thought to be an old soldier's grave. This morning was different. As she approached the grave across the beach she noticed the upper stone thrown back and the grave empty. There were many footprints about the grave, the cross had been displaced and at the bottom several silver coins glistening in the sun. The shape of large jars still showed an impression in the sand and it could be seen where the large jars had been rolled to the shore and signs of a boat having lodged in the sand. Two further coins were found at the edge of the grave perhaps overlooked in the dark.

There were many stories over the years of pirates leaving treasure trove upon the islands and many came by sea to search. Henriette Verlaque told her daughter Elisca the following story from her childhood.

The valuable foundations.

In 1839 a ship sailed into a sheltered harbour of an outlying island and anchored for the afternoon. During the early evening a boat put ashore and the crew, of dubious origin, walked to a planter's house near the shore. They requested permission to dig on the property but the planter explained that the property was not his and that it would take several days to contact the owner so authorisation was impossible. The crew repeated their request and also offered to share any gain with the planter but he became obstructive. With that the pirates put a gun to his head and ordered him inside the house where he was guarded along with his wife and children.

The pirates measured the ground following a map in their possession and eventually started digging around one edge of the house foundation. At some depth some very large clay pots were brought to the surface and due to their weight had to be rolled down the slopes to the beach with one man on each side.

When the planter related the story to his friends expecting some sympathy for his predicament they told him he was a fool for not accepting a share of the spoils...

The Three Musketeers.

Maud's uncles, James, Henri and Dartagnon Verlaque were working during a storm to clear a tree which had fallen across a road. The Governor of Seychelles came along and his coach was stopped and it was explained that there would be some delay. Monsieur Verlaque struck up a conversation with the Governor and as the carriage was about to continue he was asked his name. "My name is Dartagnon", he replied, to which the Governor shouted back and with a wave of his hand he said "And is this Artos and Portos?"

Polishing floors.

Maud remembers watching timber floors being laid in the houses. There were no machine tools as there are today and all timber was sawn by hand and finished off rather primitively with an adze. In the evening it was traditional to hold a party on a new floor with the ultimate aim of smoothing it down. Everyone would arrive. Those that could afford it wore shoes and some with their hobnailed boots left over from previous Army campaigns. Many did not own shoes. After a musical evening with square dancing, waltz and fox-trot which lasted well into the night the floor would be left smoothened down.

The next day came the finishing touches. A red-brown stain was produced from mangroves and applied to the floor. Basic wax would be smoothed on and the helpers would buff the floor with one foot placed in a dried coconut husk. With a twisting action across the front and to the side and a twist of the body Maud thought it may have been the origin of the sega dance. It certainly appeared so...

Moutia.

When the slaves were brought from Africa they were very valuable workers and the plantation owners wanted to ensure their health or valuable work would be lost. If the men fought amongst themselves it was often fatal so if caught fighting they were severely punished. If an owner came along those fighting would pretend to be dancing and he would pass by happily unaware of what was happening. It is understood that this is the origin of this movement which is danced as far away as Brazil...

CHAPTER 9

SAVY DESCENDENTS.

DeWet Savy was born in 1902, the son of Marie Elisca Savy and Henri Verlaque. This was the year the second Boer War finished and he was named after the great Boer soldier and game hunter Christian DeWet (1854-1922).

When DeWet was taken along for his innoculations as a small boy to the English Doctor Bradley he sat bravely without a murmur for his treatment. Dr Bradley told him he was fearless like the man he was named after. With that remark DeWet pointed his finger like a gun and pretended to shoot the doctor. Dr Bradley laughed saying, "Now we have a Boer shooting an Englishman..."

Dr Bradley was Chief Medical Officer and well loved. He stayed in Seychelles until he died...

DeWet Savy also married into the Ducrotoy family when he married Julia the youngest daughter of Philibert and Julia Ducrotoy. Her life is described elsewhere in this book. DeWet died between 1968 and 1973.

Other children of Marie Elisca Savy were:

Celio Cebert Savy born 1895. He died in Seychelles aged seventy five years.

Henry Furcy Savy born 1897. Married Edma Ducrotoy and died in Kenya.

Carl Ronald Savy, married but had no children and died of hepatitis in Seychelles.

Alienne Regina Savy born 1909 married Mr Gauntier and had twin daughters. She died between 1945 and 1951.

Maurice Francois Savy born 1898 and died in 1982. Maurice, it is said, was a bit backward in learning and he would go to school with trousers the wrong size or back to front and shoes too large. He found a ring one day outside the church and took it as a heavenly sign that he would marry one day and said he would keep it for his wife. Of course the children laughed at him but he did eventually marry. He obtained the money for a new pair of shoes and to give him some measure of independence he was sent to the shop alone.He came home with such large shoes meant for a man while he was still a boy. He was asked why he bought such large shoes that he couldn't walk.

Maurice replied, "The Chinaman tried to cheat me. He offered me

small shoes. for the same price I could buy large shoes".
Mauricia Therese Savy married Tusaint Hoareau a fisherman in Seychelles. She died around 1983.
Louis Filis Savy died 1st August 1947. His wife Agnes nee Nanty died in 1970.
LOUIS FILIS SAVY the son of Marie Elisca Savy born on 23rd July 1905 married Baugerine Kerbella Nanty (born 21 October 1906) in Seychelles 4th January 1932. There were five children, Carl, Lise, Georgette, Vivienne and Myra.
This is a suitable place to tell the remarkable story of how we met up with these cousins on a South Pacific island thereby putting the family in touch once again. Rene had never met his cousins as they were brought up in Seychelles. We were living on the remote island of New Hebrides, it was a weekend and in the process of breadmaking having run out of yeast with the shops closed. It suddenly occurred to me to visit the baker and buy some of his supplies. As we stood there Rene met an official from the British Office, Neil Bull, who had recently arrived on the island to act as Labour Officer and in the process, as his first case, of sorting out a work contract for Rene. Neil introduced his wife's sister-in-law from Australia and found that it was his cousin Lise, nee Savy. The irony of the situation was that if the relationship to Neil Bull, of which we had previously been unaware, become known, our very delicate negotiations through the Court would have been jeopardised. We met in secret at each other's houses under cover of darkness........we won the Court case...
Lise Serena Savy, daughter of policeman Louis Filis Savy and Bergelrine (Agnes) Nanty, born 27 April 1934, a seamstress, married on 24th June 1961. to Guy Marcel Pothin of Seychelles, born 5th March 1932, a welder. They have three children.
Yan Louis Pothin born 30 September 1962, fashion designer in Sydney, Australia.
Gary Bernard Pothin born 17 April 1964, a screen printer in Sydney, married Kerry Louise Ratray (born 20 December 1962).Their children are Harley Gene Pothin born 22 November 1985 and Joel Patrick Pothin born 29 July 1987.
Daphne Renee Pothin born 6th February 1966, an electronic technician in Sydney.

Georgette Savy the daughter of Louis Filis Savy and Agnes Nanty, born 23 May 1937, married Joseph André, now living in Perth, Australia.
Children: Caroline André, born 11.4.1958 (had Mathew and Melanie).
June André, born 14 June 1959 (had Angelique and George).
Maria André, 27 June 1960, psychologist, had Nadia and Monique).
Fred André, born 27th June 1961.
Robert André, born 8 December 1962 (had Jessie).
Jacques André, born 30 April 1964 (had Joshua and Mark).
Tina André, born 29 April 1965.

***

Carl Ronald Savy, son of Louis Filis Savy. Married and had many children in the Seychelles. He was a clever man with electronics and invented various products. He worked in England for Decca and then for Phillips in Holland, later returning to Seychelles to run his own business. When ships arrived at Mahe he would be called to fix the electronic gear. On one occasion a ship berthed with radar problems and this was situated up high and only accessible by ladder. Being frightened of heights he asked the crew to bring it to the deck where he could work. They told him this was impossible and asked him to stand on a platform with an enclosed cage for safety. Again he declined. One of the ship's officers, not relishing the idea of staying in port with a disabled radar for an indefinite period, called to the crew to bring a bottle of whiskey. Some hours later, and feeling much more confident he managed to climb to the antenna and finish the job... The family described Carl as having a pleasant nature with blonde hair.

Vivienne Savy, born 1938, the daughter of Louis Filis Savy, married Norman André in Seychelles and had six children. She married her second husband Michael Whiteside at Lady of Lourdes, Portsmouth, England on 17 11.1990. Vivienne and Norman's children:

-Marie Antoinette Linda André born in Nairobi 22 5 1957. She married 28.11.1976 at Evangelical Church, Coventry to David

Raymond Hodges. They have Maria Leslie Hodges,born Nuneaton,22 August 1986. Craig Peter Hodges, born Perth,Australia on 28 November 1988. Shana Rosemary Vivienne Hodges born Perth, W.A.,4th May 1992. They are presently living at Beechborough,Perth,Australia.

-Valerie Maggie André born in Nairobi, married Stephen James Fryer on 4th July 1992 at St Johns Roman Catholic Cathedral, Portsmouth. They have Mathew James Fryer born 28 September 1992 and Monique born July 1994.

Vernon Terence André, born in Nairobi on 11 April 1960.

Leslie Joseph André born in Tanganika on 19 March 1961. He was ordained a priest of the Roman Catholic Church at St Josephs, Mill Hill, London on 27th May 1989. He officiated at the wedding of his sisters Valerie and Anita. At the time of writing he has been assigned to the island of Antique in the Phillipines.

Anita Pauline André born in Coventry, England on 9th November 1963.

Roger Louis André born in Coventry, England on 18 February 1966.

***

Myra Marie-Therese Savy, born on 14th January 1933 in Seychelles, the daughter of Louis Filis Savy (1905-1947) and Agnes Nanty. Myra married Francois Ah Man in 1948 and she became a widow in 1968. Myra cared for many of her elders, the family looking to her for support. She came to England for an operation, on returning to Seychelles she decided to visit England again where she met her second husband, also from Seychelles, who was working in London. He is Joseph Lionel Buisson, son of Louis Buisson. Joseph is also a descendent of the Edmond family on the maternal line.

Myra and her first husband had eight children:

-Kathleen Percine Ah Man, born 8 April 1953.married and living in with two children in New Zealand.1956.

-Percy Joseph Louis Ah Man. He has three children and living in Seychelles.

-Cecile Marie Louise Ah Man born Seychelles 18 June 1957. She married Mr Berlouis, they have two children and living in Seychelles.

-Eveline Marie Claire Ah Man born 8.7.1958. Married Mr Stravens, have son Stephen Arthur Stravens born 1975.All born Seychelles.

-Daniel Victor Ah Man born in Seychelles 25th March 1961.

-Claude Pierre Ah Man born Seychelles, 29th September 1962.

-Agnes Maria Ah Man born Seychelles, 13th January 1964. Married, living in England with two children, Ella born 1990, Carol Nina 1992.

-Sybil Janine Ah Man born Seychelles, 7th January 1966.

Children of Philibert Ducrotoy and Marie Julia Edmond...

CHAPTER 10

PHILABERT OLIVIER JULEFY DUCROTOY
1896 - 1954

Born Mahé, Seychelles.

THE DEBONAIR BACHELOR

Philibert Olivier Julefy Ducrotoy II, the third but first surviving son of Philibert and Marie Julia Ducrotoy of Seychelles and later Kenya, born in Seychelles on 14th January 1896. Educated on Mahe, he worked in his father's coach-building business.

In 1912 Philibert was travelling across Mahe to make a delivery in his father's coach when ahead of him he noticed an old man on the road ahead. He was approaching a narrow pass so slowed the horses. The man was drunk and instead of walking forward he clambered up the steep bank to avoid the carriage. Philibert hastened the horses onwards and as passed the man suddenly lost his footing and fell under the carriage wheels. He died the same day of his injuries.

Several years later Philibert followed his father, who had enlisted nine months before, into the East African Army in the First World War and set off for Kenya in November 1915. He served for two years in the Transport Corps. During this time he quarrelled with an Officer and hit him with his fist. For this he was severely punished and hung up by his thumbs and left hanging. Eventually they broke and he suffered for the rest of his life with a deformity of the hand.

He enjoyed his bachelor existence but since he was by then the only son, grandma, anxious for him to settle down and have a family and not one to sit around waiting she booked her passage back to Seychelles to find him suitable wife. She had warned him of her intentions, and he agreed that he had reached an age of settling down threw up the challenge, "You find me a wife and I'll marry her."

Grandmama set sail on the next steamer and nothing was heard for one month. A letter arrived in Kenya to announce that she had found a suitable girl.

Maud Savy meanwhile had been introduced to grandmama Ducrotoy on Mahe and her mother, who was then struggling alone, advised her to take the opportunity of settling in Kenya. Maud was to travel to Kenya to meet Philibert before making a decision but, as she told her family years later, it would have been embarrassing to return to Seychelles without a husband. Whatever she said it might be construed that the prospective bridegroom had rejected her. Luckily Philibert and Maud were happy with each other. Maud's brother Henry had married Philibert's sister, Edma. Julia Ducrotoy later married DeWet Savy.

Marriage in Mombasa.

Maud and Philibert were married on the ninth of February 1924 at the church of the Catholic Mission, Seydie District of Mombasa. The marriage certificate states she was eighteen years but in fact the ceremony was a month before her eighteenth birthday. Philibert gives his father's occupation as farmer.

At the time of marriage Philibert was foreman at Polling construction contractors for Mombasa Harbour. He had been trained as a blacksmith but since he was educated he was able to take charge of the crew. This was a time of prosperity for the family but lean years were to follow as the Depression approached.

When the Harbour was completed at Mombasa the family spent a short time at Miwani with children Rita and Albert, then took the train to work at Arusha Sugar and Sizle Estate.The family were startled the first evening to hear lions roaring as they were on the edge of the National Park but the servant reassured them that it was a great distance away.

The family's time at Arusha was cut short when Philibert suffered from pneumonia and spent a month in hospital and penicillin was not then available. It must be remembered that if the man of the house did not work there was no income or a house to live in since this accompanied the job so it was a worrying time.

Philibert had over the years built his own house with supplies as they could be afforded and became available and this was at Makupa in Mombasa. As soon as Philibert was well enough to travel it was to the Makupa house to which they returned. Grandmama Ducrotoy was already living there.

After a period of recuperation the family moved to a sugar estate at Ramisi which is thirty six miles south of Mombasa and many years were spent here.

It was in 1940 that DeWet Savy returned to give Philibert news of a gold mine paying well. The two men journeyed back to Kisumu on Lake Victoria by train and then made their way to Rostermann Gold Mine in Kakamega. Philibert worked for six months before the family were able to follow on...

CHAPTER 11

GOODS TRAIN TO KISUMU.

In 1940 Philibert employed at a gold mine in Kakamega sent home the housekeeping money for the family in Mombassa. These were hard times and it was agreed that it would be more economical for Maude and the eight children to join him. It was not long after the world depression and since money was scarce Maude was forced to take the third class carriage on the train. There was the train which carried passengers and mail but third class was a carriage hooked onto the goods train. Rene was just ten years old and remembers setting off at four in the afternoon from Mombasa.

It was a long journey and two nights were spent with no heating and frequent and irritating stops for delivery of goods at every station. Luckily since there were nine family members they were able to curtain off a cubicle with a blanket for some privacy. Nevertheless, the pervading odour of the Masai tribesmen and their goats and chickens drifted along the carriage.The guard gave them hot water from the engine to make tea.

At seven the next morning the train arrived at Nairobi where sister-in-law Bella and her children came to visit them as the train waited in the station.

When Bella saw Maude in the third class carriage she climbed into the carriage in tears crying, "How lowly we have become since the old days..." This remark of course did nothing to soothe the family's discomfort.

The train continued on its arduous schedule across Kenya and twenty four hours later they arrived in Kisumu. Philibert was waiting for his family and he was soon joined by cousin Philip who helped load the truck. The truck was already fully loaded with sacks of sugar, flour and onions and their cases were also placed aboard. The nine children clambered upon the back each seeking the best place to rest on the sacks. Then the trouble started...

It was wartime and it was the rule that a serviceman in uniform could hitch a ride with any vehicle. A man from the King's African Rifles came to the truck and demanded a lift back to his village. It was politely explained that there wasn't room but he refused to listen and insisted on accompanying them. Phillip, known for his temper, lashed out and struck the soldier who in return took his cane and hit Phillip back. This infuriated him and he punched the soldier the result being a bloody nose and ruined uniform. The children watched with a mixture of bewilderment and fear as the events unfolded. The Indians who knew Phillip told him to leave quickly and shortly after an official from the nearby Army office came to investigate.

It was generally agreed that a man had attacked the soldier but neither Philibert nor the the Indian shop owners had 'seen the man before...a complete stranger'. The children sat on their sacks too frightened to speak. The soldier spent the rest of the day at the first-aid post and the four hour journey lay ahead of them so they set off without him. They reached Kakamega in the late afternoon gorged on the food prepared by the servant. It was the first time they had lived in a reed mat house constructed of a wooden frame to which woven reed mats were tacked. The houses were popular with the white ants and with a tendency to rot after a long fall of rain. There was a tin roof which was deafening during the heavy rains. Both Stella and Nancy the two youngest children were born here.

Phillip, sadly, finally met his match one dark night twelve years later. He came to stay with Leon and family at Kwamjumvu with his wife and family. Uncle DeWet Savy from Seychelles had come to stay along with wife Julia and the children. With the servants the household was bursting at the foundations. Phillip had driven back from town and en route his car collided with a truck and the rear window smashed. It was moonlight and he could see the Arab driver and in rage he set upon him. In self defence the Arab pulled out a knife and almost severed Phillip's right arm above the elbow. He managed to drive home and collapsed into bed beside his wife. At midnight the household were wakened by a scream as Phillip's wife found the bed saturated with blood. There was no telephone for miles so an attempt was made to stop the flow of blood and although still conscious he was in bad shape. It was too late to save him and he died before reaching hospital.

Papa's finger in the sugar cane.

Philibert came through the track upon a tractor on the sugar cane estate one afternoon and spotted a train rail under great pressure. With the men he tried to force it back all the time thinking that if it should spring back it would hit him on the head for certain. The rail suddenly crashed down and landed on Philibert's finger cutting it off at the first joint. Both the finger and Philibert were rushed off to hospital. Unfortunately it was impossible to save it and the doctor just stitched the end and discharged him. As he left the doctor reverently handed him a dish with the amputated finger to take home.

Philibert was brought home later that evening by Ensilm, a friend from South Africa, the finger bandaged and bloody. They were both drunk and Ensilm with his arm around Philibert's shoulder said, "Look at my poor friend."

The next morning Philibert arose early and threw his amputated finger in the sugar cane field beside the house. He received 800 shillings compensation which was about forty pounds sterling. This would represent approximately four hundred pounds in today's value.

Death on motor-cycle.

Philibert died tragically in a motor cycle accident at Kisumu in 1954. The family considered he was on his way home from work and should have been covered by insurance. For some reason he was on a different road to his usual route, only he knew why, and his employers refused to pay up. He collided with a truck at a crossroad and sustained a head injury, not regaining consciousness. He is buried at Kisumu.

Simba.

For many years the family had a servant called Simba and he was a favourite with all the children who loved him dearly and he adored them. On pay day Simba would set off to the store and return later that day with many bags of assorted sweets which he distributed between the children. It was a form of insurance. Each day the children would secrete a slice of bread, a chicken leg from the dinner table and take it to Simba. It was a welcome addition to the food already supplied by Maude. Halfway through the month Simba would run short of money so Maud would give him a shilling to spend and suggest that he should not spend his money on the children but every pay day it was the same...

Mombasa lions. During a long drought in Mombasa the animals were starving and coming closer to civilisation for water and food. One night the servant was awakened at 3am and hearing a grunt which he believed to be a donkey he took a stick with which to chase it. He rushed to the end of the garden and leapt through the undergrowth and near some bushes he came face to face with a lion...

His shouts woke most of the town and Mr Kahn an Indian and professional hunter was called upon to shoot it.

Intruder in the night...

One sultry night, Maud, awakened by the sound of someone in her room in Mombasa saw the figure of a man in the moonlight. Her vision was obscured by the mosquito net drapes which hung about her tightly tucked in. Thinking it was her son Albert who often awakened in the night when he was ill she called out, "What is the matter Albert?"

To her dismay Albert called out from his distant bedroom so she knew the figure before her was not her son.

Maud called again, "What do you want?"

Instead of running the figure slowly moved away and disappeared into the darkness...

There was no gun in the house though the doors were always locked and the windows had grilles upon them, nevertheless, a sleepless night followed. Rest only came when the servant arrived to make breakfast.

The delectable stew...

Maud came into the kitchen of the Kakamega house and asked the cook to fetch a parcel from a neighbour some distance away. There was no-one else to make the journey. The servant of all tasks being ill in his room so the cook had to go. The cook, in the process of preparing vegetables for the steaming stew seemed reluctant to depart and expressed the wish to continue cooking. Maud, puzzled by this sudden display of industry, explained that she would tend the bubbling pots and, if need be, serve the food to the family. The cook hung his head, twisted his foot on the earth floor and then looked away. Maud by this time becoming impatient shouted, "Go". The cook unenthusiastically sloped off...

The hungry family were seated around the table as Maud spooned out the stew into serving bowls. Out came something long, rather like spaghetti, and then one of the daughters screamed, "It looks like the intestines of a chicken..." It was... Maud investigated further with the spoon wondering what other delicacy she would chance upon and out came two chickens feet complete with claws. At the bottom of the pot was a chicken head, eyes tightly shut, sodden feathers dripping with gravy and a bright red comb...

Unbeknown to the family the cook had been cooking his own dinner simultaneously thus taking advantage of all the spices and herbs to add flavour to his precious titbits, and just before serving removed the tasty morsels to his own bowl. The family had been enjoying his particularly delicious stews for many months...

The servant causes chaos.

The day the servant brought home news is hilarious to relate yet at the time quite a dramatic event.

Maud gave birth to Willie, a premature baby of only seven months gestation. He was very weakly and grandmama took charge, but soon after Maud became dangerously ill with septicaemia and she was admitted to hospital. There was a long convalescence in the ward and the family at home were busy with raising the children, seeing to the men, meals to prepare and the new delicate baby to care for. All the aunts were pressed into service with a variety of tasks. Maud grew tired of the hospital food so each day the servant was dispatched with a tasty meal and he returned with the empty plate.

Maud said to him one day, thinking the family at home would be interested in some news, "Tell the family at home that the old lady in the next bed died in the night."

The servant trotted home and announced to all assembled, "The Mrs has died..."

With that grandmama and the sisters became hysterical. Grandmama wrung her hands as she weeped and said, "How can I tell Maud's parents in Seychelles that their dear, young daughter has died?" She quickly sent someone to break the bad news to husband Philibert who was by that time at work.

The servant was alarmed, indeed nervous at the reaction his news had created even though he was aware that they knew the old woman in Maud's ward.

Grandmama, in the best French tradition, had started to plan the funeral by laying out her mourning clothes when a call came from Octavia in the kitchen. Octavia had unwrapped the basket the servant had brought from the hospital and retorted. "If Maud is dead then *who* ate the food we sent?"

The servant shifted from foot to foot looking at the floor and whispered, "Mrs with the baby..."

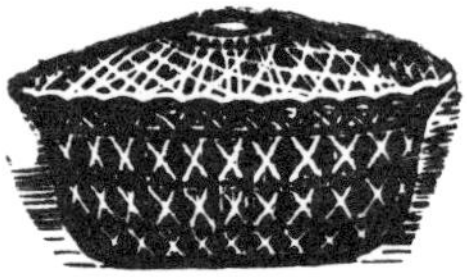

It is difficult to imagine nowadays the difficulties encountered then. Few people had telephones, the cars were old and unreliable, the roads, at best, were atrocious, the doctor and hospital could be many miles distant. René remembers a journey from Kisii to Lumbwa taking seven days by truck. Many miles of mud, having to pay tribesmen to push the truck through mud as it became bogged and having to make frequent stops for refreshments. Seven days, yet that trip now takes just two hours...

CHAPTER 12

STRANGE ANIMALS IN KENYA.

One morning in 1950 DeWet Savy and his brother-in-law Philibert Ducrotoy were walking to work early one morning when beside the road they came across part of the carcass of an animal. It was the size of a rabbit or large rat, partly destroyed by animals and the ants had eaten the head but from the top of the skull there were two distinct horns. Now animals with horns have hooves but when examined the feet had paws complete with claws. It was mystifying for them both and they called over some local tribesmen well acquainted with wildlife and they were equally astonished...

Several years ago this article appeared in the Daily Telegraph in London. (Reuters,1981). Quite a different animal mentioned but it demonstrates that there are still discoveries to be made. There were rumours in the past that there was an animal which had zebra hindquarters and the front like an antelope with a large black tongue. Zoologists have now recorded the okapi though it is still rare.

The animal found by DeWet and Philibert on the Kakamega roadside has still to be identified...

The couple in the Congo incident claimed to have seen a dark grey creature with a neck eighteen inches wide and a smaller head. The couple were in a boat and saw the creature from about thirty yards distance. The twenty porters on the expedition confirmed a sighting. Photographs taken of the creature at the time were not available at the press conference

'DINOSAUR' SIGHTED IN CONGO

A California couple say they have seen and possibly photographed a dinosaur-like creature similar to one described in Congolese native tales for centuries.

Mr Herman Regusters and his wife Kim flew into Brazzaville, the caital of the Republic of the Congo, on Dec. 9, to seek "Mokele-Mbembe"—a beast described by natives for at least 200 years and thought by some to be a survivor of a dinosaur species believed extinct for 60 million years.

They travelled to the Lake Tele area, 500 miles inland, where they spotted and photographed a creature "dark brownish in colour, skin appeared slick and smooth, long neck, small head, snakelike." Mr Reguster, 47, said the animal was "not identifiable to native hunters, Congolese forest rangers or us."

The doubting newsmen and television cameramen wrily commented that the only prehistoric monster seen was an emblem depicting a dinosaur on the couple's shirts.

When the Queen slept at the bottom of the garden.

There have been many Royal visits to Kenya over the years. Maud and Philibert recall the day in 1928 when they dressed in their Sunday clothes, suit, hat and gloves and waited near the Makupa Bridge to wave to the Prince of Wales, later King Edward VIII. He stood on a platform near the engine, waving and smiling as the train glided down to Chamgamwe Station. His hunting safari was abandoned when the condition of his father, King George V, worsened.

Maud, a Royalist, obtained a superb vantage point to wave to the Queen Mother in 1960. The local Indian shopkeeper in Tudor lived in an apartment over his shop. He invited the family to stand on his large balcony, the only one overhanging the road. As the Queen Mother passed in the open car she looked up and waved to them, to Maud's delight.

It is amusing to relate that in 1952 the Queen and Duke of Edinburgh spent the night at the end of René's garden at Kitale. René travelled home from work one afternoon and on leaving the main road he approached the rail crossing on a smaller dusty road. A member of the Special Police stopped him and asked him where he was going. René replied, "I'm going home. I live here."

They allowed him to pass. The rear garden, planted with maize, bordered on the railway track and the Royal train settled there for the night. The track at that point ran between a gap in the hillside very private, apart from René, and a suitable vantage point for security.The Royal couple travelled on to Treetops and abandoned their visit on receiving the tragic news that George VI had died...

The Ducrotoy family at Kakamega (except Winnie who had died at Ramisi). The man at rear left is a friend Roger Frichot of Seychelles.

Rita and Rivaltz's wedding 12 October 1946 Kakamega, Kenya.
From left: Nancy, Iris, Rene, Hilda (chief bridesmaid) Philibert Ducrotoy, Rita Ducrotoy the bride, Rivaltz Jorre de Saint Jorre the groom, Stella (small bridesmaid), Maude Ducrotoy, Muriel (bridesmaid) Willie and Albert.

Gunfire during a tropical night in Aden.

The time came to leave Kenya and the only chance of joining a direct passage to New Zealand was by flying direct to Aden to board the Oriana.

The week before leaving Maud had become ill and instead of improving deteriorated. They had gone to stay with Rita prior to their departure and the night before discussed the possibility of abandoning the trip but there was no return. The passports had been stamped with exits visas and that was it. The family could not call a doctor to see Maud for fear that he would prevent her from travelling. She rallied forth the following day, though weak, and with Albert and René flew to Aden. With the fervency of their travel plans it had escaped their notice that there was an uprising in Aden and as they descended the aircraft steps they were photographed. A security man with a cine-camera took particular notice of René and followed him with the zoom lens until they reached the arrival shed.

Instead of the top hotels on the seafront, and since they had a few days to spend they chose to go to the Crater, the Arab quarter, where the Arab taxi driver knew a cheaper hotel with food. As they travelled the five miles they noticed British soldiers lining the streets and intersections as they passed. They enjoyed an evening meal and the only item on the menu that Maud fancied was strawberries and the Italian waiter brought them with a shrug of his shoulders. Now, strawberries are grown in Kenya and not particularly expensive there but of course in Aden all the food is flown in. When the bill came the strawberries were three times the price of a main meal, for one small serving.

The second shock came through the night. There was an enormous boom and then gunfire further down the street and mystified René looked out of the window, there was nothing. When he was questioned the next day the Arab said,"Oh, a bit of fighting is quite normal."

The British soldiers were exchanging fire with the Arabs.

It was only later on reaching New Zealand and reading the news that they discovered they stayed in the middle of a war zone.

They wiled away the time searching for a suit for Albert. Whenever they found coat to fit, the trousers were too big and vice versa. He eventually ended up with a mismatched suit when they ran out of time.

The travel agent had also failed to inform them that the Oriana anchored offshore and passengers, after travelling the choppy sea by motor boat had to climb the precarious steps up the side of the ship to embark. Maud, still feeling weak from her illness had to jump onto the pontoon with the crew to catch her. The boat heaved with the swell and each time the boat rose up Albert and René encouraged her to jump before the boat dropped lower than the pontoon.She then decided it would be easier to return to Kenya. René said, "We can't go back. There is nowhere to go back to. Besides there is a war on in town!" With that she attempted an almighty leap which no-one was expecting and landed on top of the weedy seaman almost flattening him. The sea voyage was a convalescence for Maud and when they reached Australia some old friends from Seychelles came aboard to visit.

Maud and Albert continued living in New Zealand until 1971 and then joined her daughter Nancy and Philip Savy in Brisbane for three years.

Maud Vivienne Ducrotoy, after ailing some months, died on the 18th July 1974 at the age of sixty eight years. She was buried at Pinnaroo Lawn Cemetery, Aspley in Brisbane. In the family tradition there was just one quirk of fate at this time.

In the New Hebrides I collected the post and the telegram stated that Maud had died. The following day we received a telegram to say that Maud was dangerously ill. Since the telegrams were not dated it just left doubt in our minds whether a mistake had been made...

> **DUCROTOY**, Mrs. Maud Vivianne, of 15 Joalah Crescent, Ferny Hills (late of Auckland, N.Z.).
>
> The Relatives and Friends of Mr. Philibert Ducrotoy (deceased), Mr. and Mrs. R. St. Jorre and Family (Perth), Mr. A. Ducrotoy (Ferny Hills), Mr. and Mrs. H. Hoareau and Family (Kisii Kenya), Mr. and Mrs. R. Ducrotoy and Family (Vila, New Hebrides), Mr. and Mrs. W. Ducrotoy and Family (Melbourne) Mr. and Mrs. W. Soffientini and Family (Asmara Ethopia), Mr. and Mrs. H. G. Gow and Family (Newdeer Scotland), Mr. and Mrs. P. Savy and Family (Ferny Hills), and Mr. and Mrs. B. Muir and Family (Auckland, N.Z.), are respectfully advised that the Funeral of his beloved Wife, their Mother, Mother-in-law, Grandmother and Great-grandmother is appointed to leave St. Joan of Arc Church, Clyde Road, Herston, Today (Friday) for Pinnaroo Lawn Resting Place, Aspley on conclusion of Requiem Mass commencing at 2.15 p.m.
>
> METROPOLITAN FUNERALS,
> Valley and Mt. Gravatt.

Now that the Savy family have been chronicled the story continues with the descendants of Philibert Ducrotoy and Marie Julia Edmund and start with their son Joseph who, at such a young age, came to a tragic end...

CHAPTER 13

THE BOLTING HORSE.

Joseph Ducrotoy born in Seychelles, the second son of Philibert and Marie Julia Ducrotoy, died most tragically as a youth.

He came with the family to Kenya and worked as a stable boy for Lady DeCoville and lived over the stable. Philibert came to visit him on furlough one afternoon and reported back to the family that he was extremely happy having his own room, bed and all his possessions about him.

When the two Philiberts joined up for the first world war Joseph had been greatly disappointed as he had plans to accompany them. He was far too young so Grandmama bought him a small uniform and pith helmet in which he marched up and down.

Joseph had been teaching his sisters to sing a birthday song for their father and told them to pick flowers to spring a surprise. It was the traditional French song:

"Mon petit Papa,
C'est aujourd'hui ta fete..."

The song is supposed to be a happy song but when I first heard my daughter sing it on returning from her French school, being unaware, at the time, of the family story, I remarked, "Oh, what a sad song...". Yet it is only poignant for our family.

Joseph set off in the horse and carriage on his father's birthday and as he reached a bend in the road a truck sped by. For a bit of fun, the Arab truck driver having one of those new horns beeped it loudly as he passed. The horse reared in fright and Joseph fell ahead into the road where the wheels passed over the lower part of his body as the horse bolted. Joseph was taken home and he stayed conscious for some hours, talking until he died. The day was his father's birthday, August 12th , and Joseph not yet sixteen...

Julia Ducrotoy, daughter and last child of Philibert Ducrotoy and Marie Julia Edmond, was born in February 1910 or 1911. Born in Seychelles she attended school there before the family left for Kenya where she continued her schooling. Both Julia and her sister Paquerette were taught lace-making by the French nuns and became proficient in this work. Julia married Maud Savy's brother DeWet James Savy and seven or eight years elapsed before she had her first child. The following children were born in quick succession. Neville Savy now thought to be in Australia. Nadege Savy married an Indian and emigrated to Australia. Rose Savy married James Ernestine in Kenya and now lives in Australia. Patrick Savy. Frederick Savy returned to Seychelles. Julia Savy born in 1952 married in Australia.

Julia, warned that her heart would not stand the strain of having many children, disregarded this advice her desire for children being so great. The inevitable happened and Julia died in childbirth when her last child was born. She was buried in Mombasa Cemetery, Kenya.

The baby, named Julia after her mother, went to live with her aunt Edma, Julia's sister. Young Julia was raised by her and eventually emigrated with her aunt to Melbourne, Australia.

The nuns at Kifungito Orphanage took the remaining five children for several years as DeWet their father had to work but he visited as often as possible. After a year or two DeWet wrote to his sister Mauricia in Seychelles asking her to find him a suitable wife. A wife to help him and take on seven children...no easy task. She wasted no time and very soon a letter was received by DeWet with the news that prospective bride had been found. No other details...

Since DeWet was by this time fifty five years old he envisaged an older woman who would be a companion and housekeeper. Imagine his astonishment and delight when Nataleine arrived by ship and just eighteen years old. She was barely older than her charges but she settled down happily. Unfortunately Nataleine lost a child and there were no further children of the marriage. They both returned to the Seychelles in 1968 and DeWet died there in 1973. Nataleine's brother was a policeman in Seychelles.

Children and grandchildren of Ducrotoy and Edmond of Seychelles.

1.The grave of Julia Savy, nee Ducrotoy, at Mombasa Cemetery in December 1954. 2. Savy children at Kifungito Catholic Orphanage.

Paquerette consults a fortune teller...

Delightful stories of Paquerette's eccentricity had reached me over the years and it was no disappointment to meet her for the first time in 1990 several years before her death.

Maria Emily Paquerette Ducrotoy, the daughter of Philibert and Julia, was born in Seychelles on 11th April 1906. She was always known as Paquerette. She attended school in Seychelles but completed her education in Kenya so she did not speak French as well as her sisters.

She married Claude Guillamin, a Frenchman from the West of France. During the Abyssinian War he joined up with the Italian Army but as a cook not a conscript. He was an excellent cook and made his living selling cakes, sweets, drink and food to the soldiers.

Paquerette, during his absence, went to visit a fortune teller, a Mauritian woman called Mrs Pitts. The cards were read foretelling Guillamin's death and Mrs Pitts told Paquerette she could see that she was a widow. Paquerette laughed saying "My husband is not in the fighting force of the war... besides, I received a letter from him today..."

Some weeks later Paquerette was acutely distressed to receive a letter of condolence from her mother and father-in-law in France and sympathy at the loss of her husband and their son. The Italian authorities had, in error,, written only to his parents and on hearing the shocking news they assumed she had already been informed.

Paquerette had no children. She went out for some years with the director of a French shipping line but she never married again. She had a dressmaking shop in Mombasa but in 1967 followed her sister to Melbourne, Australia where she stayed for the rest of her life. She was always a great source of amusement when the family visited her. She walked about with a huge bunch of keys and had numerous boxes one inside the other all locked and each having its own key.

Paquerette announces her own death...

Willie and Josephine visited Paquerette regularly even though she lived a distance away. One dark, cold evening they returned from work after a long day to find a telegram which read, "Paquerette died yesterday." They were greatly shocked and although they were exhausted they set off on public transport, the arduous journey across Melbourne to her house.

They tapped at the wooden door and were startled when Paquerette answered the door to them... She grumbled that she hadn't seen anything of the family for so long and she knew that a telegram would bring some visitors...

Most of her spare time was spent suing a bus company which she blamed for her injuries when she toppled off the back of a bus in Melbourne. When she won the case she then sued the solicitor for failing to reach a higher figure...

When Rita's daughter, Daphne, was married in Perth they assumed, thankfully, that it would be too great a distance for her to put in an appearance. Paquerette informed the family of her flight to Perth and someone drew the short straw to fetch her. There was a big delay since she could not ascend the steps of the aircraft and eventually she reached the arrival hall. To Paquerette's amusement a very blonde porter pushed her into the airport and her first words to her family, spoken in Swahili, were, "My little white donkey is pushing my wheelchair".

The wedding went ahead without incident much to the family's surprise and then the wedding photographs arrived. It happened that Aunt Paquerette for some unknown reason had failed to feature in any of the group pictures. She was furious again. It was quietly explained that the most of the guests had dispersed, the wedding suits back in the hire shop, besides, the honeymoon couple had departed to some unknown distant destination. This did not concern Aunt Paquerette who insisted she would not fly back to Melbourne until more photographs were taken.

The family, faced with the prospect though they loved her dearly, of Aunt Paquerette staying longer, indeed, perhaps indefinitely, went into action. Someone fetched the couple back from honeymoon, wedding clothes were donned yet again, flowers in place, and Aunt Paquerette taking pride of place. Honour was satisfied.

The haunted house...

Paquerette stayed in a Mombasa house with her nephew Tim who was then attending school.

After settling in the servant nervously came to Paquerette and told her there was the ghost of a man in the house. He had seen the ghost of a young man in a kanzu several times around the house as he attended to his tasks.

Several weeks later as Paquerette sat at her dressing table mirror brushing her long black hair she saw the reflection of a man in long white robes. She immediately dropped everything and rushed about the house screaming and Tim joined the servant in the search of the darkened rooms. Nothing was found initially and then Tim noticed out of the corner of his eye a figure in white near the sofa... With that they all rushed from the house and along the streets and battered on brother Philibert's door for sanctuary.

(Paquerette 1986)

PAQUERETTE'S WEDDING.

The Seychellois wedding held in Mombasa during 1925. The photograph shown on the following page was taken at the Makupa house owned by grandpapa. It was built by grandpapa Philibert on his arrival from Seychelles. The front double doors were shuttered (behind the bride) and there are windows of trellis behind the overgrown ivy on each side of the door. The trellis work, to allow air to pass, was made by grandpapa with left over barrel hoops which explains their slight curves. There is another full photograph of the house in a previous chapter. (Chapter 2)

This interesting photograph shown on the next page shows the following:

From left to right: Henriette Ducrotoy, daughter of Philibert Ducrotoy,and beside her, her husband Joseph Henri Ernestine already married in Seychelles on 2 August 1911; the mother of the bride Julia Ducrotoy (nee Edmond of Mahe); Octavia Ducrotoy; Claude Guillamin the groom; Marie Emily Paquerette Ducrotoy the bride; Philibert Ducrotoy (jr.) brother of bride; his wife Maude Ducrotoy nee Savy; Henry Furcy Savy, brother of Maude, and Henry Savy's wife Edma, nee Ducrotoy.

The three children in front: Irene Antoin (no relation, friend of the servant who insisted on being in the picture and thereby ensuring her place in history); Leonie Frost the daughter of Bella (nee Ducrotoy) Frost; Julia Ducrotoy sister of the bride who later married DeWet Savy, the brother of Maude and Henry Savy.

Note: Maude, Henry and DeWet Savy of Seychelles all married into the Ducrotoy family. Maude had already married Philibert Ducrotoy when the following photograph was taken. Henry Savy married Edma Ducrotoy. DeWet Savy married Julia Ducrotoy (one of the bridesmaids) much later.

Marriage of Paquerette Ducrotoy 1925

Edma Ducrotoy born in Seychelles the daughter of Philibert Ducrotoy and Julia Edmond attended school at Mahé and later in Kenya. Edma married Maud's brother, Henry Furcy Savy. Henry journeyed away to work and joined Bella and Joe (Hoareau) at Wanjoi, Aberdare's Mountain fifty miles from Nairobi. He was above 8,000 feet altitude unaware that he suffered from hypertension. The strain being too much he suddenly developed severe headaches which went on for several weeks before he died.

Joe Hoareau, shocked by this tragedy, sent a telegram to the family stating that Henry had died. Telegrams were hand-written in those days and on reading it the family thought that one of the children had died since one of them had been ill. The household overcome with sadness started to receive friends with condolences as the sad news spread. When an aunt arrived and read the telegram she queried, "The name is not clear. It looks like *Henry*." All assembled were bewildered. It was decided to consult the writer of the telegram to decipher the script. To speak to him they had to climb up the high steps of the Telegraph Office. The officer read out the telegram with the message that Henry had died. Edma fainted with the shock of the news and fell down the stairs to the bottom soon after miscarrying a male infant. There were no other children of the marriage. Henry Savy was buried at Wanjoi and on a later pilgrimage his brother DeWet made a concrete headstone for his grave.

Edma Ducrotoy, later married a Francis Cambra, a Portuguese Seychellois and they had four children, Frank Cambra born 1932 has three children. Martha Cambra born 1934 married James Stitfall an Englishman now living in Wellington, New Zealand. We visited them in 1970 and they had a house overlooking the sea with a son and daughter. Emily born in 1927 married an Englishman, John, with two children. Alice Cambra born 1928 also married an Englishman in the Army at Aden. They met when he came to Kenya on furlough. They have one son in the Army. Edma visited Martha in New Zealand in December 1982.

When Willie and Josephine Ducrotoy, René's brother, came to visit in England during 1984, they received a telegram from their son to state that Aunt Edma had died on 2nd October. The telegram arrived on the 5th October but by that time we had departed to travel across Europe to Milan to see sister Iris Ducrotoy-Soffientini in that city. It awaited our return.

By this time the funeral had taken place but what followed was really one of those incidents that are unique to the Ducrotoy family. Several days following Edam's funeral several officials, cap in hand, arrived to speak to the next-of-kin and to apologise. It seemed that the funeral that they had attended was in vain since the body in the coffin was not that of Aunt Edma...

The daughter of another family had called at the funeral home to view her mother and stated that it did not look like her own mother. The official said that in death people do look different but the woman was adamant. The woman she had viewed was Aunt Edma. The mistake was then realised but our family had already attended the funeral of her mother, dispatched forthwith. A second funeral was attended by our family and this time with Edma in attendance. The funeral expenses,which had been rather costly, were waived...

Edma and Octavia's children: Emily and Alice Cambra, Gabriel Grandcourt, Frank Cambra, Lena , Margo, Lucy and Joseph Grandcourt.

Olga Hoareau and her husband Willie Sedgewick.

Philibert Hoareau and his wife Emily in 1944.

Octavia Ducrotoy born in Seychelles in May about 1907, the eighth child of Philibert Ducrotoy and Julia Edmond of Mahe. She married Lionel Grandcourt and had five children. These were Lena, Margo now in Belgium, Gabriel presently in Melbourne, Australia, Joe and Lucy Grandcourt. Octavia died in August 1989 in Melbourne, Australia.

Lena and Margo at Kelosa, 27 February 1940

Henriette Ducrotoy born in April 1894, the daughter of Philibert Ducrotoy and Marie Josephe Julia Edmond of Mahe, married Henry Ernestine of Seychelles. The following children were born, Henry; Louis; Phillip; Therese (Tess); Joseph; and Leon. Henriette died aged sixty four in Mombasa. Henry Ernestine her husband, then went to Seychelles where they had property. This would have been about the year 1965.Henry died in his eightieth year. Phillip Ernestine died around 1952 in Kenya. Joseph Ernestine died in 1989 and Louis in 1987. Leon stays in Perth, Australia and Tess in Dorset, England.

Emily and Alice Cambra daughters of Edma Ducrotoy.

Bella Ducrotoy born in April 1892 at Mahe, Seychelles, the daughter of Philibert Ducrotoy and Marie Julia Edmond. She married Mr Frost, a blacksmith and locksmith from the island of Reunion and had one daughter. Leonie Frost married Mark Godley in Kenya and went to

South Africa. They produced four sons, Ogilvy,Maxim, Carl and Derek. Ogilvy married in Kenya and is now living in Australia. Carl married and has sons in Australia. Derek became ill and was operated on in South Africa for pressure on the skull and made a good recovery. Unfortunately, some time later he was hit by a truck there and killed. Leonie, Mark and Carl Godley are also in Australia.

Bella parted from Mr Frost and later married Joseph Hoareau and had eight children:

-Louis Hoareau. René met him once only when he arrived on board ship at the Port of Mombasa. He was travelling from Tanzania en route to Seychelles , looking very handsome and very much like Clark Gable in looks. Louis died in Seychelles.

-Olga Hoareau married Willie Sedgewick, a fitter and turner, they are living in Australia. They did not have any offspring.

-Robert Hoareau died in Kenya around the age of twenty six years.

-Philibert Hoareau married Emily from the Seychelles.

-Joseph Hoareau, a very handsome man, bought a farm in Kenya and became a Freemason. He had two children, then he married a Miss Clark from Kenya.He spent a year in England and came back with an English wife.

-Joseph died in Eldoret in his mid-fifties.

-Helen Hoareau married an Englishman, Stan Tyler.

-Beatrice Hoareau married an Englishman and came to England.

-Thomas Hoareau was in the Kenyan Army during the Second World War and demobbed in Kenya in the late fifties. There was an increased immigration of Italians into Kenya after the war, as there were to many parts of the Southern Hemisphere, and they filled many posts, in fact there were many advertisements stating "Italian preferred" since they were known to be extremely hard workers. Thomas had an Italian foreman where he worked. The boss called him to the office one day and asked him why he refused to take orders from the Italian. Thomas stated that during the recent war he had sworn to King and country that he would fight the Germans and Italians and he would uphold that, therefore why should he take orders from them now... Thomas later went to the Congo where he spoke French fluently, he married a Seychellois woman in Kenya and had several children. One of the sons is a pilot with Air Seychelles at the present time...

These are the children and grandchildren of Bella nee Ducrotoy who died in Kenya in 1965.

Octavia at Dar Es Salaam circa 1945.
Octavia nee Ducrotoy and children Lena, Margo, Gabriel and Lucy Grandcourt.

Jeanne Marie Adelie Ducrotoy born in Seychelles on 3rd February 1900 the daughter of Philibert Ducrotoy and Marie Julia Edmond. Jeanne married Louis Samuel Morgan, they had an only child named Emmanuel.

Emmanuel married Theresa Ernestine daughter of Henriette nee Ducrotoy and Henry Ernestine in Nairobi.

They had five children, Patricia, Elizabeth, Judith, Christine and David. Rene and I with the children visited Jeanne on a trip to England en route to Brazil in 1975. She was then in hospital and very frail but told us about her stay in the village of Vignacourt, being well-acquainted with the family remaining there.

Monique Deray one of the French cousins told us that Jeanne had lived in the village of Vignacourt for a period of time. One day a party was to be held for Serge Ducrotoy, the elder of the family. He was highly thought of, a Magistrate of the Court and Major in the French Army. He had one fault though that of his time-keeping and he failed to put in an appearance. Jeanne became impatient with waiting, and despite Monique saying they *must* wait for him, declared, since she was one of the elders of the party, that the party should begin. And begin it did...

Jeanne Morgan (nee Ducrotoy) died at the age of eighty four years on 16th January 1984 and is buried in Gillingham, Dorset.

Emmanuel and Therese Morgan are now living in England.

We continue now with the descendents of Philibert Ducrotoy jr. and Maud Savy of Seychelles and Kenya.

CHAPTER 14

THE LOVELORN APPRENTICE.

Albert Reginald Ducrotoy born in Seychelles on the 26th May 1928, the first son of Philibert and Maud Ducrotoy. Maud travelled to visit her family in Seychelles by steamship and Albert was born there. There was great rejoicing when a son was born and much celebrating by his father Philibert. However, as the months passed it became obvious there was something wrong. His development, physically and intellectually, was impaired and this slowly became evident. His grandmama Ducrotoy and other relatives played an active part in teaching repeatedly, to walk, to dress himself and eventually to read. In fact he reads everything in sight. Any script and important letters left lying around the house are memorised for future regurgitation. There is an amusing example of this when many years ago he almost started a local scandal.

A workmate of René's, walking out with a very nice girl, approached René one day during the coffee break. He confided that he was seeing a girl more educated than himself and was ashamed since he was illiterate. He begged René to write a love letter to her as he dictated the words so she would not know. René detested writing and the thought of even one letter was a daunting prospect however he did have some sympathy for the love-struck lad. The letter went off and back came a reply. Of course this placed René in the position where he had to write a second one. The letters continued with more and more passion as the courtship continued and René found his spare time taken up with more and more letters...

Albert went most days to fetch the mail from the Sugar Estate Office where the bundles of post were left on the main desk. One day he noticed an envelope in René's handwriting addressed to the girl. By this time René and his uneducated workmate had been transferred a hundred miles away to Molo. Albert, interested that his brother was courting, and excited that he should be the sole bearer of news to a village where one could dine out for weeks on any snippet of gossip, rushed home...

Albert imparted the information to his incredulous mother and although she told Albert he must be mistaken, nonetheless she knew Albert's skill in recognising handwriting. Now in those days one did not write formally to a girl unless there was some serious courting afoot so there was much speculation. Furthermore, Maud was aggrieved that she had not been taken into René's confidence. It was some weeks before René had leave and came home to explain his actions...astonished and amused at the family uproar he, and Albert, had inadvertently caused...

Albert speaks three languages, Swahili, English and Creole and has the most remarkable memory. He actually helped me write the entire present day family records recently in Brisbane, Australia. None of the family were sure of all the birthdates but Albert knew the dates, the names and occupations, not only the immediate family but their wives and husbands as well. Subsequent searches of official documents have proved his accuracy. Recently he overheard me trying to remember the year I had entered Australia eighteen years before. Albert told me the month and the day of the week and old passport entries later proved this. While his long term memory is remarkable his short term is poor in that he has to have a shopping list for a long list of items. The local shopkeepers chuckle and say Albert is the only one in Australia to obtain credit. Since he cannot estimate the cost before setting off from home he may be one dollar short at the checkout but the shops know he will be back the following day. Although Albert reads so well he is only able to cope with signing his name. It is fortunate that ball-point pens were invented, for prior to that, he pressed so hard every signature of his name cost the family the price of a nib as they were discarded.

Today Albert lives in Brisbane with his sister Nancy and Philip Savy.

Rita Julia Ducrotoy, the daughter of Philibert Ducrotoy and Maude Savy, born in Kenya 2nd January 1925. Rita, well educated at the White Sisters Convent School in Mombasa, stayed on at the Convent as a teacher. There for several years and during that time teaching, amongst others, her own brothers and sisters. Therese Ernestine, a cousin, also taught at the school. As René recalls, "Having a sister and cousin as teachers, we as children had expectation of favours but we didn't get any..." Rita left when the family transferred to Rosterman Gold Mines at Kakamega and she opened her own school for the mining children there. Rita married Rivaltz Jorre de St Jorre who was born in Seychelles on 9 April 1919, the son of Mahé Jorre de St Jorre, an old Seychellois family. The ceremony took place on 12 October 1946 at the Catholic church, Kakamega, Kenya. The seven children:
Edward Philibert France Jorre' born 6 January 1949 in Nairobi. He

WHITE SISTERS CONVENT SCHOOL 1941

Back row: Teacher from Goa; RITA DUCROTOY; Bertha Clark; Odette Clark; Portuguese teacher; Elsie Clark and Millie Bartrum.
Front row: Sister Modesta; Sister Mary; Mother Superior(Theresa); Sister Hermina and Sister Mary; name unknown.

married Wendy Baker on 14 January 1978 in Perth. They have two children Aaron John Edward Jorre' born 7 May 1981, and Trina born 10 February 1983 at Perth, Australia.
Marjorie Jorre' born 15 October 1949 at 7am. The child lived just twelve hours and died at 7pm.
Norman Roger Adolphe Jorre' born at Kitale, 12 August 1952. He married Paula McDonald of New Zealand on 14th February 1976. Their two daughters are Chantalle Yvette Jorre' born 1 March 1982 and Angelique born 16 October 1987. All living in Perth, Australia.
Daphne Dorothy Dawn Jorre' born at Kitale, 13 March 1955. Married Martin Osborn 15 April 1978 at Queen of Martyrs Catholic Church, Maylands, Perth. Martin is from England and has been working at Mount Newman Mines, Western Australia for many years.
Sarah Angela Valerie Jorre' born Kitale 12 5 1959. Air hostess for Ansett Air Australia. Married Michael Morrisey, a pilot, on 10 4.1982.
Brian David Julian Jorre' born 25 March 1951 at Kitale, Kenya.Married Suzanne, sister of Paula (Norman's wife) on 16 April 1977. They have two children: Hayley born 20 December 1978 and Brendon David Jorre' born 14 February 1981 in Australia.
Marc Rivaltz Mahẽ Jorre de St Jorre born at Nakuru, Kenya on 19 November 1963.

Winnie Blanche Ducrotoy the daughter of Philibert Ducrotoy and Maud Savy, born on 19th September 1937. She was a lovely child and ran, laughed and played with the other children. She became suddenly ill one evening and died.

René remembers her being ill and complaining of a headache. The doctor came and gave medication but to no avail. She died at Ramisi.
Maud had returned prior to this from helping to lay out a dead friend. Grandmama Ducrotoy scolded Maud for this since it was an old Seychellois custom or rule never to attend a dead person and come straight to a child, for the sake of the child's health. Of course, we know now that Winnie died of cerebral malaria which causes many deaths to this day.

Her godfather Ronald Savy paid for the tombstone which should still be in place today in the Catholic cemetery at Mombassa. René and Willie returned on a nostalgic trip many years later and it took them one and a half hours to drive there. Yet years before René remembers the mourning family setting out at seven in the morning and not

RITA DUCROTOY as teacher in Standard 1, White Sisters Convent.
Apart from Rita (centre front), in the second row from the front, third from left is Hilda Ducrotoy aged 9 years. Third row from the front is Albert Ducrotoy aged 13 years, he is third from left.

White Sisters Convent School 1941, now Star of the Sea School.

A SEYCHELLOIS FAMILY 1949.

The Ducrotoy family 1949. Back row: Rita, Rene, Albert, Willie, and Hilda.
Front row: Muriel, Stella, Maude and Philibert Ducrotoy, Nancy, and Iris.

reaching Mombasa ferry until six in the evening. The transport was then poor. Old trucks, weak tyres prone to puncturing and the incessant mud for endless miles. There were a few Indian shops along the roadside where a stop could be made for tea, bread and sardines...

Hilda Ducrotoy and Edmond Hoareau

Hilda (nee Ducrotoy) and Edmond Hoareau with baby Henry (who later died), Olivier and Richard at Kisumu in 1953. Edmond was the son of Albert and Elia Hoareau from Seychelles and Kenya.

Hilda Elisca Ducrotoy was born on 5th August 1932 the daughter of Philibert and Maud Ducrotoy. Her official birth certificate states she was born on the 5th September that year but she was born the month before. The parents were late registering the baby's birth so not wanting to be fined for late registration they stated she was born a month later.

Hilda married Edmond Hoareau of Seychelles on 19th February 1949 at St Joseph's Catholic Church, Kisumu, Kenya. They had nine

children : Oliver Terence Edmond Hoareau at Kisumu on 25 January 1950. He married Catherine Edmond. He married Catherine Edmond in Australia. He married his second wife Sharon McFhee and their son Fabian was born 19th December 1986.
Henry Stephen Hoareau born Kisumu on 15th June 1953. Sadly, Henry died of cerebral malaria at the age of three years at Kisumu.
Peter Joachin Hoareau born 9th April 1956 at Kisumu. Married Dawn Anglum on 9th March 1985 at St Jinbarrs Catholic Church, Ashgrove, Brisbane.
Julian Benjamin Hoareau born 10 November 1961 at Kisumu. Married Carol Jennifer Elizabeth on 15 December 1984 at Our Lady of Victories Catholic Church, Bowen Hills, Brisbane. Their daughter Jaqueline Hoareau was born 27 September 1986.
Simon James Etienne Hoareau born 22 November 1962 at Kisii, Kenya. Married Kelly Lee brown on 5 March 1988 at Brisbane.
Cecile Mary Janet Hoareau born 1 August 1954 at Kisumu, Kenya. Married Tryggve Salter on 25 November 1989 at Queens Apostles, Stafford, Brisbane. They have two children Amanda born 30 December 1980 and Keir Edmond Salter born 24 July 1990, 8.15am a premature baby and now a strapping lad.
Anne-Marie Hoareau born Kisumu, Kenya on 29 September 1957. Married Les Vaagsbo Salter, the brother of Tryggve (Cecile's husband). They have two daughters, Bianca Erika Salter 15 June 1984 and Michelle Jade Salter 20 November 1987.
Lucille Agnes Therese Hoareau born Kisumu 14 January 1960. Married Henricus Vanzanten on 23 August 1980 at St Ambrose Church, Brisbane. Second husband is Paul Grainger an Englishman. They have Andrew born 31 August 1983, Mathew born 13 February 1987 and Christopher born 12 January 1990.
Richard Christopher Hoareau born at Kisii on 9 March 1951. His first wife was Robin Noumann. His second marriage is to Robyn Yvonne Perry on 27 October 1990. Children are Wayne born 24 May 1970 and Nicole born Christmas Day 1972.

Hilda and Edmond Hoareau stayed on in Kenya after other family members had migrated. They followed many years later and now live in Brisbane, Australia.

Ducrotoy family of Seychelles and Kenya continued...

Hilda Ducrotoy sitting beside the hot water system. Servants lit a fire beneath the tank and the hot water would flow into the house.

Muriel Ducrotoy

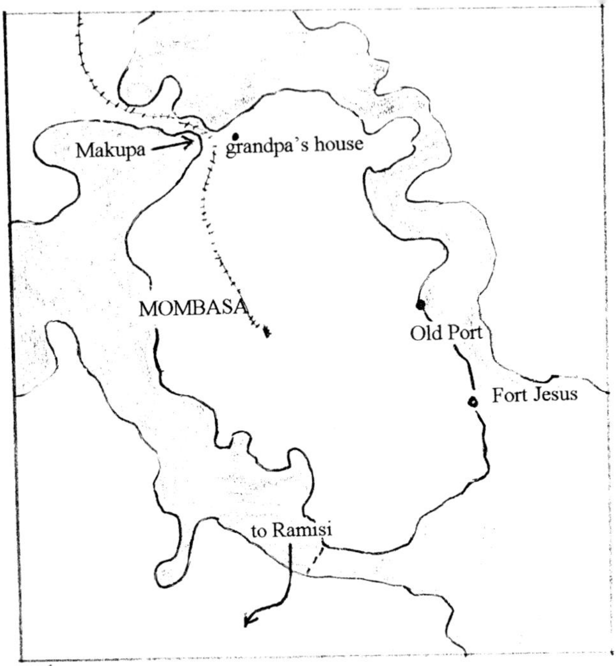

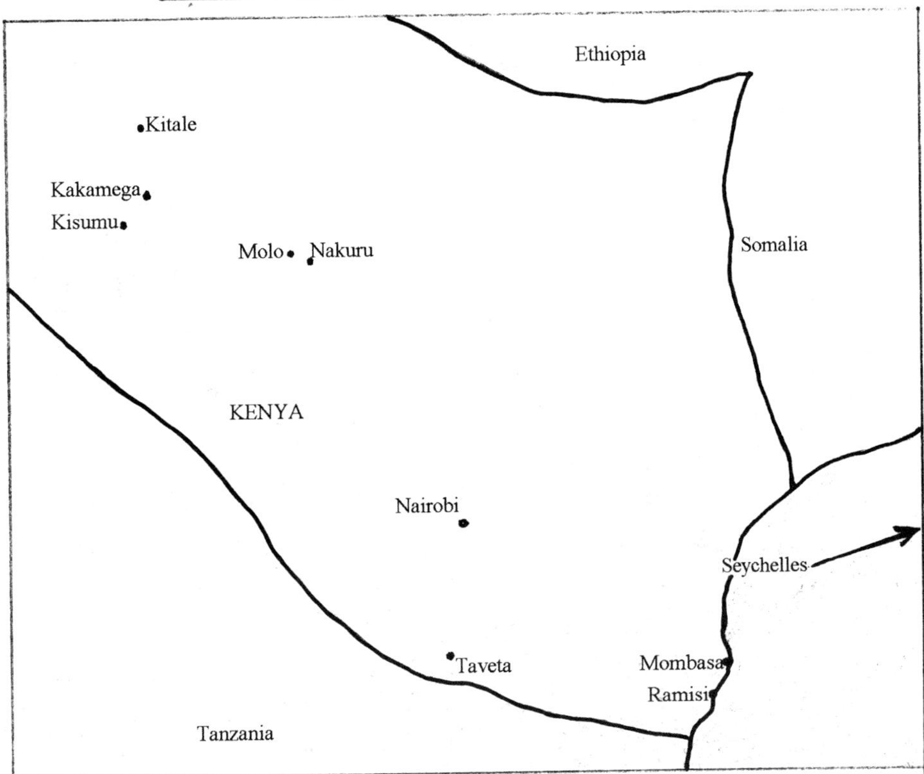

Map of places mentioned in this book.

Willie, Rita, Nancy, Rene and Hilda Ducrotoy at Molo, Kenya. 1949.

MAXSIM RENÉ DUCROTOY

Born Mombasa, Kenya. 1930

Maxsim René Ducrotoy, the third son of Philibert and Maud Ducrotoy, born at Makupa in Mombassa, Kenya on 3rd November 1930. Attended White Sisters Convent School, Mombassa and the White Fathers College at Yala.

Starving at Yala

It was decided to send René and Willie away to school and Philibert considered his son's education would be complete if he sent them to the Marist College at Yala. His friend's children, the Andre's, were already at the school besides he schooled under the Marist Brothers in Seychelles.The Marist Brothers were from Switzerland and France. The Bishop transferred them all to Uganda and sent in a group of White Fathers from Holland and England just before René and Willie were enrolled. René remembers some speaking Dutch and of course this was wartime 1941. Our relative Father Leslie has mentioned though that the College at Yala having been administered by the Mill Hill Mission in London.

The boys welcomed the move as an adventure being the first time away from home although it was just thirty kilometers away. Their friends Joseph, Francois and Norman André had already been with the College for a year.

The school huts where they slept had dried cow dung floors which was the traditional method of building though the mission had a brick kiln. This led to the boys suffering from jiggers but this was the least of their worries. They were starving...

The boys had their own cook and he was supplied with the raw food and this should have been quite sufficient except that he provided food to all his family within the local village and cut the boys portions. René and Willie, along with the other boys, started to get thin and pale and the Father Superior came to their hut one evening suspecting they were homesick and wasting away.

"What is wrong with you boys, you were so rosy cheeked when you arrived?" he remarked.

They dared not say that the cook was stealing the food in case he cut off their supply of sustenance in spite. They kept silent.

René had four shillings and Willie had twelve shillings and whenever the strict regime allowed they managed to walk to the nearest Indian shop at a small village close to the school. They bought peanuts and chocolate until the funds ran out . The money that was to last them the whole term lasted several weeks only.

The regime was :

7am Church
8am Porridge with lard and salt-
pepper upon it. Cup of tea.
8.30am Classes.
12pm Beef or fish, rice and onions served in a
sticky stew.
1pm. Classes
3pm. Afternoon tea, one bun.
3-4pm. Classes.
4-6pm. Sport.
6pm. Prayers and hymns.
7pm. Leftovers of the main meal.
8-9pm. Classes.

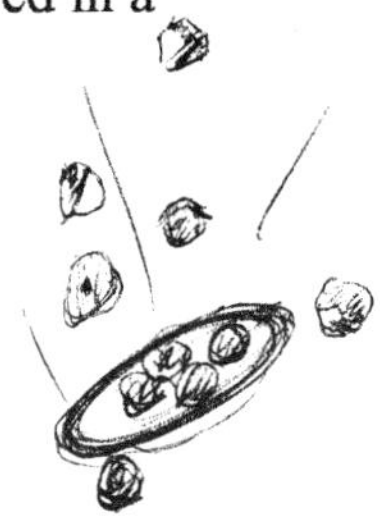

This was for six days a week. Sunday was free except for morning and evening mass in church. Once a week agriculture classes were held. Every boy would take a hoe and work tilling the fields though the boys suspected it was a method of farming their vast estate.

On special days for a treat they had pigs ears served in a sour salty sauce considered a delicacy.

The boys favourite time was afternoon tea. The door of the serving hatch to the kitchen would be opened and a tray of buns placed there for them all. They listened each afternoon around three o'clock and the ones with the best hearing seemed to get the biggest buns as they raced ahead to grab them. The boys behind would knock the tray into the air and grab them as they tumbled down. They were eaten long before the tea was poured.

At the end of term Mr André called at the school for his sons and to save Willie and René from waiting until the weekend when transport would be provided he offered them a lift home for the evening and they would continue the journey the next morning. Mrs André was amazed when the boys ate all the food on the table.

When Willie and René eventually reached home with their pinched faces and long hair,Grandmama wrung her hands and cried, "How thin you are and you have wasted away to nothing. That settles it. You are not going back to that College, they are not going to starve my boys."

These photographs have been included to illustrate the construction and various aspects of the houses. Note the grilles on windows for security, thatched rooves and water tanks for storage. Lower photograph shows René and sister Rita with her family at Kitale.

Philibert, their father, was disappointed because it was his last hope that his sons would receive a good education.

The College was informed that the boys would not be returning and they enrolled at the Indian College at Kakamega being the only school suitable. This meant a walk of three miles each way but it was twelve miles for the servant. The servant accompanied them and then returned home repeating the journey in the afternoon to fetch them. There was two hours of lessons in English and four hours of Hindustani. It was at this school that the René became interested in mathematics.

The dog in the coal box.

René, alongside brother Willie, after leaving school apprenticed at Rosterman Gold Mine at Kakamega in 1946. The name of the mine evolved by three men having founded the business. Messrs. Ross, Foster and Holman.

René and Willie worked alongside Ali, a Somali apprentice who was very thin and black with a wide grin showing a mouthful of small teeth. Ali played many tricks on his apprentice friends but he surpassed himself one hot afternoon. An amusing incident happened when the Foreman was absent.

The workshop door was a frame covered in mesh with a chain on the handle which hung below the door. If not fastened the chain tended to catch on the ground. The mesh door opened onto a courtyard where the Foreman kept his savage dog on a chain. Whenever the apprentices went to fetch tools they had to walk in a wide circle as the dog strained and growled to reach them. Mr Singh, the Foreman, was called out on an urgent call and before leaving gave the boys a task. He left Ali in charge and finally locked the dog into the coal box in the corner of the yard. This was a safety measure since the dog was so dangerous and he fastened the lid of the box with a staple. The box held coal and coal dust for the furnace. All went well until Ali thought he would have some fun. He placed the air hose into the coal box which blew black coal dust all over the unfortunate animal. Either in fear or anger the dog leapt up and broke through the folding box cover snapping the staple as he vaulted into the courtyard. Ali still showing bravado taunted the dog as he pulled the door behind him but the chain caught on the gound and wedged it ajar. The blackened furious dog took his chance and squeezed through the door and continued the chase with bared teeth.

All the apprentices scattered in terror. René and Willie jumped onto a nearby high bench but the dog was only interested in Ali.

Ali ran and ran about the workshop with torn trousers desperately trying to find a safe spot and with every step losing face as the junior apprentices laughed heartily from their safe perches.

There was little work done that afternoon. It was explained to the foreman on his return to the workshop that the dog had managed to break out of his 'secure' box. It was most fortunate that dog's cannot talk...

In Defence of Kenya..

Renê's cousin Tim joined the Kenya Regiment during World War 2. He was then aged around twenty years of age. With twelve other men and some Ascaries he was sent on patrol on the outskirts of Nairobi in the Ngong Hills looking forward to his first assignment.

His cousins, Philabert, Joe and Robert Hoareau came to hear of this and they decided on a plan to scare the living daylights out of him. With some difficulty they managed to locate the camp where the tents had been erected and sat back to wait until evening. Tim was settling down at dusk under canvas after an exhausting day on manoeuvres when he heard some rustling noises in the dry scrub beyond the ring of tents. Ready for action he decided to tackle the approaching enemy. As he keenly charged through the tent flap he stopped to grab his 303 rifle which he discharged with abandon into the bush. There was a yell and realising they were being shot at, Phil, Joe and Robert scattered like frightened rabbits. They returned home with the comfort that the enemy had no chance of invading Kenya whilst Tim was on patrol...

Haile Selassie.

In 1948 Renê went to Creightons in Molo followed by employment at Kitale in 1952. Self employed for three years he then accepted a position with Motor Mart at Mombasa, then transferred with the same firm to Tanzania. He was later granted a supervisor's position at African Oil Refinery in Mombasa.

There were many visitors to the Shell refinery and soon after starting Renê met Haile Selassie when he came on a goodwill visit. The next visitor was Kenyatta for the inauguration held after the refinery had been running two years. Renê was asked to make a model of the oil refinery as a presentation gift and he designed and made a cigarette box and lighter. He made the oil tank complete with steps running around the side as the box. The chimney with flare at the top was the lighter. This chimney also had a ladder running up the side. When finished it was copied in silver and mounted on varnished wood. During the presentation the flare of the oil refinery was turned off waiting for Kenyatta to turn it on. The speeches went on longer than expected and the gas built up and when finally lit there was quite a blast and terrific heat. René's workshop had been set up as a bar so the drinks flowed freely that afternoon. He continued at the refinery for three and a half years until departing from Kenya for New Zealand with Albert and Maude.

René and I met at Spanish language classes in Auckland and lessons floundered when we became engaged a year later. We were to regret this neglect of the Spanish language many years later when we landed (illegally as it happens) after crossing a swollen river in a remote part of Brazil. The river was the border of Bolivia and the two border policemen refused to speak Portuguese, but this is another story...

René was offered a foreman's job on the island of Espiritu Santo in the South Pacific. We were married on arrival by the British Commissioner, Chris Turner. He later received the CBE from the Queen and we often laugh and say Her Majesty must have learnt of his task. We have often thought of applying to the Guinness Book of Records for the cheapest wedding and honeymoon on record. It cost $1. That would be 30 pence in today's money or 3 Seychelles rupees. It came about this way.

We were staying in the local hotel, the old Corsica, which was burnt down many years ago. It had been converted out of an old aeroplane hangar after the last war. Red painted floors, bamboo, shutters on the windows with large ceiling fans for the intense heat. Everywhere there were tropical plants and bright green geckos running across the whitewashed walls. The hotel had an ambience that modern hotels could not recreate. It had a Somerset Maugham character about it.

The owners of the Corsica were Antoinette and Alan Brown and we asked them to be witnesses at our wedding which they accepted. Following the ceremony at the British Office we returned to the Corsica and Chris Turner ordered champagne saying, "It's the least the British Government could do for its citizens..." Antoinette and Alan Brown refused to supply us with a hotel bill although we insisted.

For our honeymoon we booked a unique trip which, as luck would have it, had been arranged to observe an ancient ceremony on Pentecost Island. It was one of the few times when outsiders were allowed to see the ceremony and we were warned before leaving that it may be cancelled at the last minute. The Melanesian men throw themselves from a high platform hundreds of feet in the air. Their ankles are tied with long vines to support the drop however as the day progressed the vines dried out and snapped. Many were unconscious as they hit the ground. *

The local entrepreneur who had arranged the small trip for us and a

* Years later the jump was copied for the sport of bungee jumping.

group of international reporters forgot to send the promised picnic basket. René and I were perfectly happy drinking sweet coconut water straight from the nut and eating lap lap with the local people. Lap lap was a mixture of unidentified meat (it doesn't pay to ask any questions) and taro covered in leaves and steamed in coconut. When the overseas photographers saw the slimy food covered in flies they went quite green. As a result of this slight hiccup the tour operator refunded our money for this extremely unique trip.

This photograph was taken of René aboard the ship 'Himalaya' during lifeboat practice with the children. We were sailing from Brisbane, Australia to New Hebrides in 1973. Simone was eight years, Melanie five, Chantal nearly two and Marcel eight months old.

René was employed in building tugboats in the Gubbay shipyard at Luganville and then for the French Government Public Works. We later transferred to the island of Efaté on New Hebrides for a four year contract with Burns Philp. This meant travelling on the French ship 'Polynesie'. All our crates of furniture, children, their nurse, were transported on the overnight voyage.

1974. Chantal, Athalie, Marcel and Rene Ducrotoy drinking green coconuts during a trip to Fila Island for the day.

Our children are: Richard presently in South America; Michael in Australia with one son Zachary Michael; James in New Zealand married to Lorraine with two children Simone and Janine; Simone in Kent with one son Samuel Christopher; Mélanie Avena Louisa in Kent with one son Aaron Joseph; Chantal Maria born Luganville 8 February 1972; Marcel Maxsim Ducrotoy born 13 March 1973 at Port Vila.

Cyclone at sea...
We sailed from New Zealand aboard 'Marconi' of the Lloyd Triestino Shipping Line and encountered a cyclone in the South Pacific. René and Marcel were the only ones to report to the dining hall for every meal quite oblivious to the fact that most of the crew sat in their life-jackets waiting in the lounges for the order to abandon ship. The weaker members of the family lay in their bunks praying, not for deliverance from the storm, but for a speedy sinking so we could be out of our misery. The ship was seriously off course and three days late entering the port of Fiji. The island of Fiji had been devastated by the cyclone and a clean-up operation under way but it was wonderful to set foot on, if not dry land, steady land once more.

Chantal ill...
The ship continued to Tahiti where a paediatrician was brought on board to treat Chantal who was seriously ill. She was so ill he came out to the ship with the harbour pilot to save time having had a radio message sent ahead. The ship's Captain planned for me to stay in Tahiti with her and fly on to rejoin the family in Mexico. Luckily she responded to treatment rapidly before the ship left port so the ship's doctor assumed responsibility once more.

Shipwreck at sea...
After Mexico we sailed through the Panama Canal, to Curacao and then the long crossing of the Atlantic. We were sitting on deck one afternoon and René mentioned that we were travelling in the wrong direction. When you are at sea with water on all sides this isn't something you concern yourself with and leave your faith with the ship's navigator. A lively argument followed with all assembled.

René said, "We should be travelling east to the Mediterranean but judging from where the sun is setting we are sailing south-west. If we continue we will arrive in South America."

Everyone present, whether Judge, plumber or dentist, no more qualified on navigation than René, all had an opinion on the subject. That 'the sun is in a different position in the tropics', that the 'world was round' and many other witless remarks and so it went on for several hours.

"Why don't you go and have a word with the Captain, maybe his navigator has nodded off." I added helpfully, hoping for some peace and quiet.

Several hours later the Captain announced that the ship had been ordered by international law to divert to the coastline of Africa to search for survivors of a shipwreck at sea. It was around five hundred miles from Casablanca where two ships, a fishing boat and a Japanese tanker had collided. One or two men had been picked up immediately but many others were missing. Late that evening we reached the area and circled for hours with other ships as the crew used searchlights on the darkened waters. There were other large passenger liners in the distance similarly engaged and it gave the night a strange and eery atmosphere. René and I, in evening dress, stood on the deck until midnight straining our eyes long after others had gone to the warmth below. All we saw was wreckage from the ships. As smaller rescue ships arrived on the scene our ship had permission to continue the following morning.

Sailing to Rio de Janeiro...

.We sailed from Genoa, aboard the Italian ship 'Augustus' bound for Rio de Janeiro. It was our intention to spend the three week voyage learning Portuguese. There were French, German and Italians aboard and by the time we arrived in Rio we could count to ten in Portuguese and had almost forgotten our native tongue...

. René accepted a position with Saveiros Camurano in shipping for Rio Port Authority in Rio de Janeiro. After several years in Rio we transferred to Jacunda, a tin mine 200 kilometres from the nearest shop at Porto Velho, Rondonia in the heart of the Amazon jungle. We travelled out once a month to stock up with supplies. René would fly to Rio three thousand miles away for a week or more at a time. The children went to the local school which was just the local church turned into a classroom. Later transferring to a diamond mine at Mato Grosso.

René crossing the Amazon River
March 1977

Philibert Olivier Julphy Willie Ducrotoy born in Mombasa on 29th June 1929 the son of Philibert Ducrotoy and Maud Savy. Willie was born prematurely and weighed less than five pounds. When he was baptised with the above names Maud's brother DeWet Savy said, "He is such a small squid. Why do you have to give him an additional burden?" Willie, since Maud was ill with septicaemia, being a weakly infant went to the care of grandmama. From that time she continued to care for the baby and he stayed with her for long periods.

Years later he accompanied Hilda and Edmond Hoareau to the Seychelles where he met his future wife. The third generation Philibert to find a wife in Seychelles... Josephine Marie Antoinette Tregarthen was born in Seychelles 24 September 1939, the daughter of Captain Antoine Tregarthen and Charlesia Bailey. Willie and Josephine were married on 27 April 1957 at Immaculate Conception Catholic Church, Victoria, Mahe, Seychelles. The same church where grandparents Philibert Ducrotoy and Julia Edmond had married sixty years before...

They have three children, Vivienne, Bernard and Jacqueline.

Vivienne Maud Margaret Ducrotoy born at Eldoret, Kenya on 23rd April 1958 married Douglas Gerald Henderson. The ceremony at St Joseph's, Springvale, Melbourne on 19 December 1977. They have Andrew Phillip Henderson 19 December 1978 and Timothy Edward Henderson 1st February 1983. Both students of St Anthony's Catholic College, Perth.

Bernard James Ducrotoy born at Kitale, Kenya 5th October 1959. Married Sandra June Robinson 28 may 1988 at Uniting Church, Dorcas Street, South Melbourne.

Jacqueline Noella Ducrotoy born at Kitale, Kenya 16 December 1961. Married Steven John Henderson (brother of Douglas) on 8 march 1980 at St Joseph's Catholic Church, Springvale, Melbourne, Australia. They have two daughters. Avalon Mary Henderson 12 August 1985 and Rebekah Catherine Henderson born 18 May 1987.

Willie and Josephine came from Kenya to Melbourne where they lived for many years and in 1993 they moved to Perth driving for several days across the Nullabor Plain. On 7th February 1976 they became Australian citizens and attended a naturalisation ceremony.

Muriel Mary Mavis Ducrotoy, daughter of Philibert and Maud Ducrotoy, born 25th March 1933.
Muriel attended the White Sisters Convent, Mombasa during the last war. She remembers the air raid practice when she was at school, the sirens going and being herded by the nuns quickly into dark cellars with just a candle for light and munching on dry biscuits to pass the time.

She worked in a grocers shop in Kitale and then chemist shop in Mombasa for some years. She met Herbert Gow, a Scotsman, who at the time was a farm manager in Kitale. The wife of the farm owner, Mrs Patterson, wrote home to Bert's mother in Scotland and told her he was planning to marry a girl from Seychelles. The family asked him to come home on holiday, thinking he would find a Scots wife and settle in Scotland. He returned to Aberdeen and whilst there he met a farmer who owned a farm in Nyasaland and he offered Bert a position as farm manager there.

Back to Nyasaland.

After a brief holiday with his parents he set off yet again to Africa. Bert drove thousands of miles to visit the farm with the purpose of accepting the position so he could marry Muriel. He took one look at the desolate farm and returned to Mombasa in despair. He then accepted employment as stevedore at Mombasa Port. The firm sent him for several months to Rotterdam. After their marriage in Mombasa, Bert took another farm manager's position at Kitale.

Defending herself against the MauMau.

Muriel was alone on the farm one evening and she sat admiring David, her new baby. The servant had retired to his quarters as darkness fell. She always had a nervous disposition but that day she had been reading of the MauMau massacres. She heard a rustling noise outside the window and anxious to defend her infant she grabbed Bert's shotgun which he kept loaded in the wardrobe. She had never used a gun in her life but she had watched the men in action. A tropical night can be dark and it was pitch black outside so she could not see anything. This small detail did not deter her and she aimed through the screens at the window to whatever was on the other side... on this occasion, a hedgehog. The force of the blast lifted her off her feet and threw her back onto the bed. She was still lying there gasping when distant neighbours, hearing the shot, came to investigate. The word soon spread that she was a lousy shot and this in itself was a deterrent to any potential marauders...

Some years later the family returned to Scotland.

Children are: David Arthur Gow born in Mombasa on 8 October 1959. He married Yvonne Robertson of Peterhead, Scotland at Old Parish Church, Peterhead on 26 September 1981. They have two sons Ryan Gow born Aberdeen 1 October 1983 and Mark Alistair Gow born Aberdeen 14 July 1986.

Heather Roseann Gow born at Kitale, Kenya on 15 January 1961. Married Raymond Summers at Old Parish Church, Peterhead on 20 September 1980. She has two children Martin Summers born at Peterhead on 18 April 1983 and Emma Logan born at Peterhead 10 June 1991.

Richard Alexander Gow born 30 December 1970 in Scotland.

Bert's death.

For some years Herbert (Bert) suffered from heart trouble and he was admitted to Aberdeen Hospital for a triple bypass in 1980. There were several odd incidents as a prelude to this which may be worth recording. When Bert and Muriel visited several months before his operation we accompanied them to Dover railway station for their departure to Scotland. Bert hung out of the train door to wave goodbye as the train glided out when I suddenly had the feeling it would be for the last time. Strangely, an odd, almost frightened look came on his face as though the thought suddenly came to him too...

As the day for the operation approached Bert discussed what Muriel should do when she was a widow, and that his daughter Heather's wedding should go ahead as planned, as though he had a premonition of his death.

The operation seemed successful but he haemorrhaged in intensive care and returned to theatre. Muriel was concerned and Rene planned to travel to Scotland to be with her the following week. The following afternoon I felt Rene should leave on that evening's train. I booked the ticket and packed Rene's bag and as he walked in from work there was a telephone call from Heather to say her father had improved and there was no need to come. However I still had the nagging feeling that Rene should go. In any case the ticket had been bought so that settled the decision. René travelled through the night and arrived in Aberdeen the next morning. Bert was sitting up and smiled when René greeted him in Swahili. René and Muriel set off to Peterhead and as they arrived René rushed to answer the telephone as they entered the house. It was the hospital to say Bert had just died. It was 15 March 1980.

Bert was buried at New Deer Cemetery, close to his father's home. In Scotland it is the custom for the men only to attend the graveside cemetery and René froze in the bitter cold wind.

Eighteen months later Muriel married Norman Robertson at the Registry Office, Peterhead on 30 October 1981. They were divorced several years later. Muriel now lives in Aberdeen.

***

Nancy Agnes Mary Ducrotoy daughter of Philibert and Maud Ducrotoy was born at Kakamega, Kenya on 7 October 1939.

Nancy married Phillip Savy from Seychelles at the Roman Catholic Church at Kitale.They continued to live at McAlder Mine after their marriage and were also on Rosterman Gold Mine. In 1968 they emigrated to Brisbane, Australia to join Philip's brother who had already entered the country. For the past twenty seven years they have resided in Ferny Hills, and have taken Australian nationality.

They have one daughter, Audrey Rina Savy born 19 October She married John D Butt the son of Dean Butt of Brisbane but formerly from Kenya, on 18th August 1984 at St Paschal's Catholic Church, Wavell Heights, Brisbane. John is with the Australian Army. They have three sons; Daniel Butt born November 1985; Nathan Butt born August 1987 and Joshua Jacob Butt born 1994. They have lived in Adelaide on Army postings but now live in Brisbane.

***

Iris Patricia Ducrotoy born in Kenya 29 November 1935, the daughter of Philibert and Maud Ducrotoy. Married Walter Antonio Soffientini from Milan. Walter's father had come to work in Abyssinia (now Ethiopia) when his boss opened another glass manufacturing plant in Kenya to which he was transferred to take charge. Walter his son joined him, travelling from Milan to Mombasa where he met Iris. They were married at St Marys Catholic Cathedral, Mombasa in 1959. After their son Angelo was born ten months later the family transferred back to the glass factory in Asmara, Abyssinia. This was in March 1961.

They had two children: Angelo Soffientini born 1 May 1960, born and baptised in Mombasa. Angelo married Laura from Genoa and has a son, Matheus (Mathew) Soffientini born September 1993. Claudia Ann Soffientini was born in Asmara, Abyssinia on 11 November 1966. Both children schooled in Abyssina and Milan. Angelo completed his military service within the Italian Navy.

After thirteen years in Asmara where Iris became fluent in Italian, the country became unsettled and during the Ethiopian uprising around 1974 the entire family rapidly departed for Milan. They had had the foresight to invest in property in Milan some years earlier and on arrival they ran an interesting coffee shop and bar serving lunches in the centre of Milan. They made one visit to England and Scotland. Walter died suddenly at the age of forty eight years of a coronary thrombosis. Nancy Savy, her sister in Brisbane, wrote a letter of condolence, dated 26 January 1986 and I quote from her letter.

> "When Josephine contacted us past midnight to give us the sad news I could not go to bed again from shock. He was still a young man, only forty eight years old. I know we all have to go some day but it's still a shock when it arrives especially when one realises how far from each other we all are..."

Walter died during one of the worst storms in European history and many flights were cancelled. René, unable to obtain a flight nor a sea crossing for Europe finally chose instead to visit Iris later in the year.

Buried in wrong grave.

The funeral was distressing enough but several weeks after the funeral the family were approached by the authorities. It appeared that Walter had been buried in the wrong grave and it was necessary to exhume Walter's body. This was disturbing but poor Angelo had to proceed with the formalities and identify the body as it was exhumed before another Catholic burial could be performed on another site...

After Walter died, Iris, to her credit, managed to continue to run the coffee shop and bar with the help of her son Angelo and some assistance from her father-in-law. The business however was sold in 1989 and today she continues to live in her own apartment in Milan.

***

Stella Maureen Dorothy Ducrotoy, daughter of Philibert and Maud Ducrotoy, their last child, born 20 May 1942 in Kakamega, Kenya. She attended school in Kitale and then Eldoret High School. On leaving school she worked in a pharmacy in Nairobi. Her photograph appeared in the newspaper, the Daily Nation, 9 April 1964. Another article from the Radio Times of Kenya dated 26 May 1960 showing her in the radio station is partly shown here.

Stella working at Radio Kenya.

Stella married Harry Brian Muir at Holy Ghost Cathedral, Mombasa on 5 August 1961.

They have two children Michelle and Yvette. Michelle was born 1 February 1963 at Arusha, baptised 7 April 1963. She gained her

Bach.Ec.Degree at Perth University on 23 April 1985. Michelle worked as an air hostess for Ansett Air, Australia. Married David, an Englishman, and now living in Perth with one son.
Yvette Muir was born in New Zealand.
Stella and Brian emigrated from Kenya during November of 1964 aboard the ship "Marconi" to join Brian's brother who had emigrated there previously and they took New Zealand citizenship. They later settled in Perth, Western Australia and became Australian citizens.

Stella Ducrotoy at Thompson Falls, Kenya in 1959.

Rules for Horse Drawn Vehicles.

Remember you are slow moving -
Keep to the left -
Stop or slow down -
Signal with your whip in plenty of time

Always sit on the offside of the cart -

By Order.

CHAPTER 15

THE ANCIENT NAME OF SAVY.

The use of modern surnames dates from the Middle Ages and French surnames came into existence around 1000AD and originally used for the nobility.

The French surname Savy is of nickname origin belonging to the group of surnames derived from a personal characteristic of the original bearer of the surname Savy originates from the old French term "savi" which means "knowing" and would have been used to identify someone of a wise and cautious nature. There are several variants of this name and these are Savy, Save and Savee. The latter being especially numerous in the south-western regions and Languedoc area of France.

The Family Crest for Savy is "Three ostrich plumes proper", (shown below). The ostrich plume denotes the nobility.

The Arms: Azure a chevron argent between three vairs or.

Specific arms were shown emblazoned on shields so those on the battlefield could be recognised.

BIRTH OF MARIE ELISCA SAVY
SEYCHELLES 1869

1869. **Birth in the ~~Colony~~ District of Seychelles.** in the Island of Mauritius.

When born and where.	Name and Surname.	Sex.	Natural or Legitimate.	Name and Surname of Father and Mother, Profession and ~~Residence~~ Domicile	Name and Surname of Informant, Profession and ~~Residence~~ Domicile	Names and Surnames, of Witnesses Profession and ~~Residence~~ Domicile	Marginal Entries.
On the fifth day of the present month of June at midnight at Port Victoria Mahe	Marie Elisca Savy	female	legitimate	Victorin Savy a joiner, and Laurestine Morel his wife both residing at Mahe	The father	Emilien Fondaumiere a proprietor and Andy Menton Fitch an usher both of age residing at Mahe.	

declared by the above said Informant in the presence of the above named Witnesses who, after I have read over this Act to them, have signed with me Officer in charge of the Civil Status of Seychelles ~~marked the same in my presence~~, this twenty second day of June One thousand ~~nine~~ eight hundred and sixty-nine with the exception of the informant who has declared his enability so to do.

Signature or mark of Informant ______

Signature or mark of Witnesses (sd) E. Fondaumiere

(sd) A.M. Fitch

(sd) R.M. Rennards.

Officer ~~of the~~ in charge of the Civil Status ~~of the Colony of Seychelles~~.

/Eighteen words erased as null.

Certified to be a true Extract:

[signature]

Officer of the Civil Status of the Colony of Seychelles.

BIDALE SAVY nee Lalanne.
DEATH AT ANSE ETOILE - MAHE - 1885

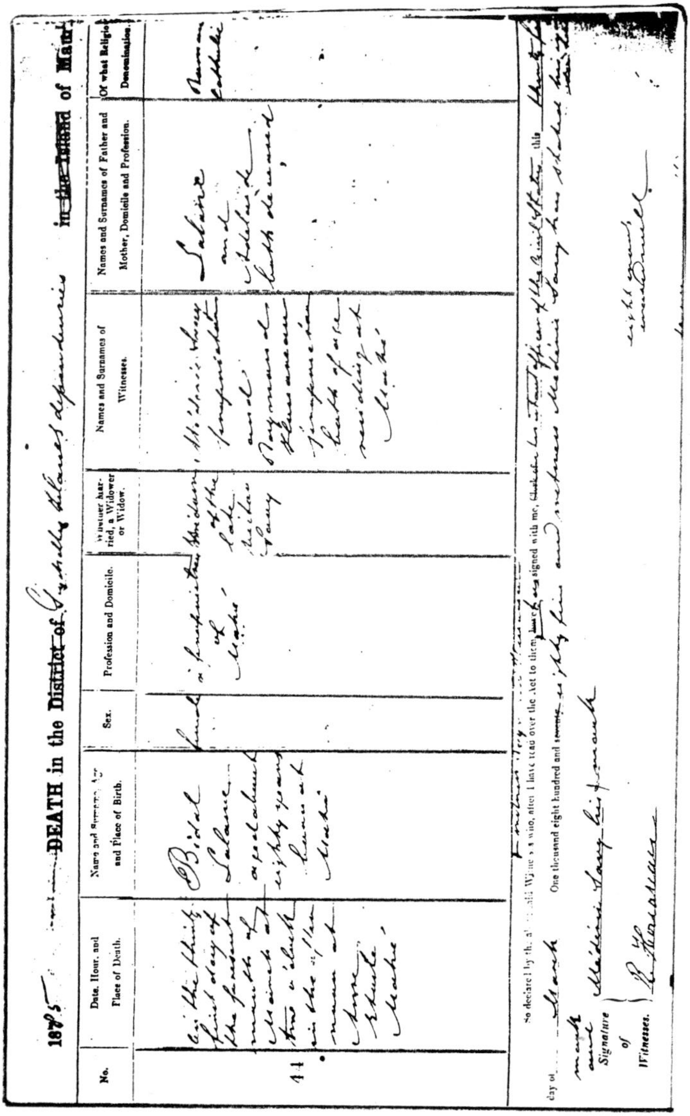

On the 31st day in the present day of March at two o'clock in the afternoon at Anse Toile. Bidale Lalanne aged about eighty years born at Mahe, a proprietress, widow of the late Victor Savy. Witness Medoric Savy proprietor and Raymond Hoareau proprietor. Parents Lalanne and Adelaide both deceased.Roman Catholic.

Official declaration by Adelaide and Lalanne 1811.

Declaration of the births of Jean Baptiste aged 6 years born 4 February 1804,Jean aged four years and eight months born 9 March 1810 and Bidale born 2 (or 10) January 1808. The natural children of Adelaide and presented by their father Lalanne.

Officially declared 22 May 1811.

47

Baptiste né le quatre février mil huit cent quatre, de Jean [illegible] le neuf mars mil huit cent dix et de Bidale, né le [illegible] janvier mil huit cent huit tous les trois Enfants naturels de la nommée Adélaïde affranchie de feu Sr Gourbel, présentement habitante domiciliée de cette Isle [illegible]

Les trois dits Enfants, nous ont été présentés et leur père reconnu en présence du sieur Emmanuel Poupinel officier de santé & Courtois dit Languerin, tous deux majeurs habitans domiciliés de cette Colonie, voisins et témoins appelés par la mère des dits Enfants.

Sur l'autorisation à nous donné par Ordonnance du Tribunal de première instance de cette isle du lundy onze février, mil huit cent onze, registré au présent registre f° 47 et verso. Les témoins ont signé avec nous, la mère ayant déclaré ne savoir signer, de ce enquis.

Poupinel

Courtois

Le Roy
Commissaire Civil

Constaté suivant la Loi, par moi François Le Roy Sous Commissaire de Marine, Administrateur Comptable & Commissaire Civil aux Isles Seychelles y faisant fonction d'officier Public de l'Etat Civil soussigné.

Le Roy

Commissariat Civil de Mahé Isles Seychelles

SEYCHELLES MARRIAGE 1865.

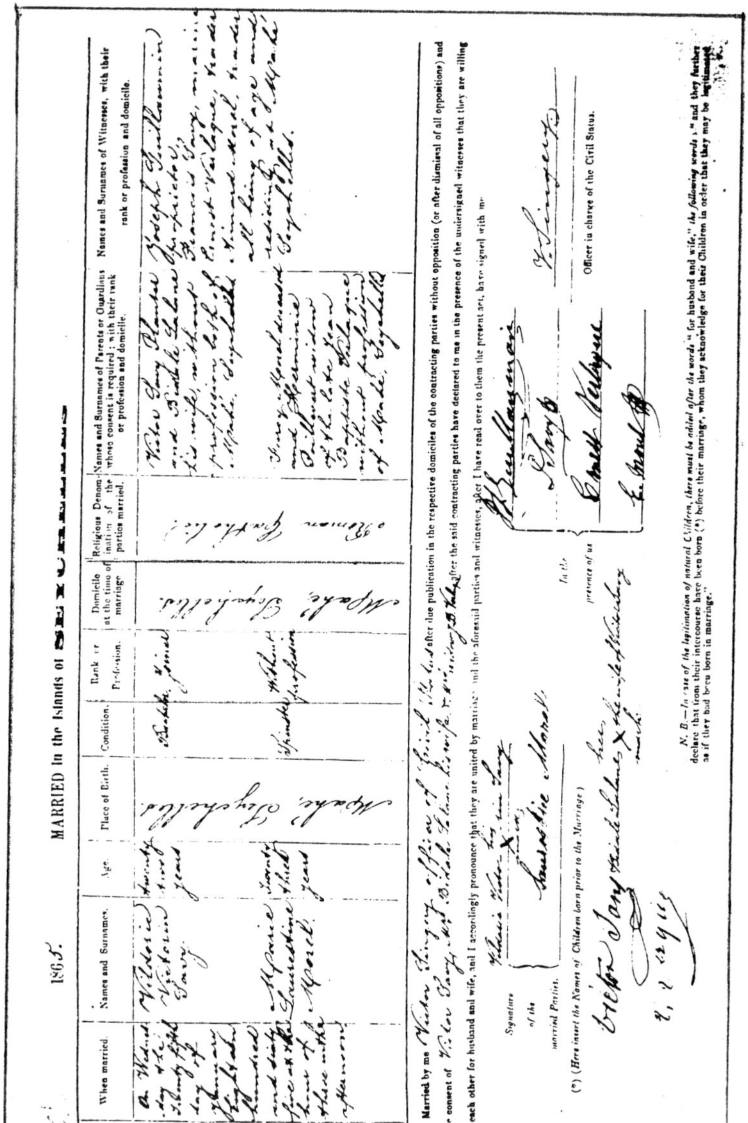

1865. MARRIED in the Islands of SEYCHELLES

No.	When married.	Names and Surnames.	Age.	Place of Birth.	Condition.	Rank or Profession.	Domicile at the time of marriage.	Religious Denomination of the parties married.	Names and Surnames of Parents or Guardians whose consent is required; with their rank or profession and domicile.	Names and Surnames of Witnesses, with their rank or profession and domicile.
1.	On Wednesday the Twenty fifth day of January Eighteen hundred and sixty five at the hour of three in the afternoon	Vildoric Victorin Savy	Twenty two years	Mahé, Seychelles	Bachelor	Joiner	Mahé, Seychelles	Roman Catholic	Victor Savy Planter and Bidale Lalane his wife, without profession both of Mahé, Seychelles	Joseph Guillaumin proprietor, Francois Savy, mariner, Ernest Vielaque, trader, Aimard Morel, trader, all being of age and residing at Mahé Seychelles.
		Marie Laurestine Morel	Twenty three years		Spinster	Without profession			Jimey Morel deceased and Herminie Pillevert widow of the late Jean Baptiste Vielaque without profession of Mahé, Seychelles	

Married by me Victor Singery officer of Civil Status after due publication in the respective domiciles of the contracting parties without opposition (or after dismissal of all oppositions) and with the consent of Victor Savy, Mrs Bidale Lalane his wife & Mrs widow J. B. Vielaque after the said contracting parties have declared to me in the presence of the undersigned witnesses that they are willing to take each other for husband and wife, and I accordingly pronounce that they are united by marriage and the aforesaid parties and witnesses, after I have read over to them the present act, have signed with me.

Signature of the married Parties. Vildoric Victorin his X mark Savy; Laurestine Morel

In the presence of us: J. Guillaumin; F. Savy; Ernest Vielaque; A. Morel

V. Singery
Officer in charge of the Civil Status.

(*) (Here insert the Names of Children born prior to the Marriage)

Victor Savy; Bidale Lalane her X mark the wife of Victor Savy

N. B.—In case of the legitimation of natural Children, there must be added after the words "for husband and wife," the following words : "and they further declare that from their intercourse have been born (*) before their marriage, whom they acknowledge for their Children in order that they may be legitimated as if they had been born in marriage."

Marriage of Vildoric Victorin Savy and Laurestine Morel of Seychelles 1865.

MARRIAGE 1891

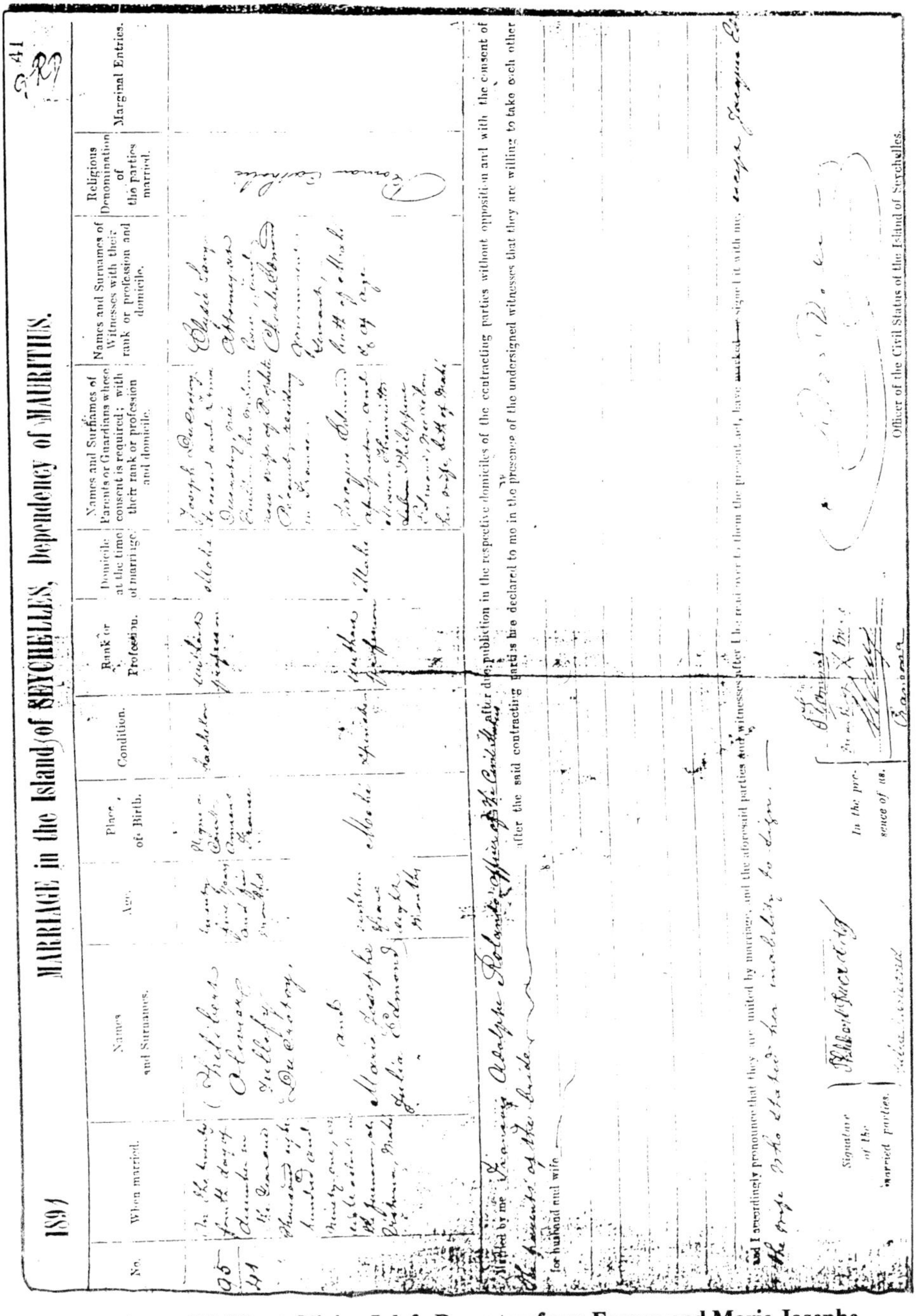

1891

MARRIAGE in the Island of SEYCHELLES, Dependency of MAURITIUS.

No.	When married.	Names and Surnames.	Age.	Place of Birth.	Condition.	Rank or Profession.	Domicile at the time of marriage.	Names and Surnames of Parents or Guardians whose consent is required; with their rank or profession and domicile.	Names and Surnames of Witnesses with their rank or profession and domicile.	Religious Denomination of the parties married.	Marginal Entries.
95 / 41	On the twenty fourth day of December in the year one thousand eight hundred and ninety one, at [illegible] in the forenoon, at Victoria, Mahé	Philibert Olivier Julefy Ducrotoy, and Marie Josephe Julia Edmond	twenty five years and five months; sixteen years eight months	[illegible] Amiens France; Mahé	bachelor; spinster	without profession; without profession	Mahé; Mahé	Joseph Ducrotoy deceased and Emma Ducrotoy, née Emilia, his widow, now wife of Raphaël Pécourt, residing in France. Jacques Edmond a proprietor, and Marie Henriette Philippine Edmond, née Lebon, his wife, both of Mahé	Elysée Savy Attorney at Law, and Charles Edmond Government Servant, both of Mahé, of age	Roman Catholic	

Married by me François Adolphe Roland, Officer of the Civil Status after due publication in the respective domiciles of the contracting parties without opposition and with the consent of the parents of the bride

after the said contracting parties have declared to me in the presence of the undersigned witnesses that they are willing to take each other for husband and wife

and I accordingly pronounce that they are united by marriage, and the aforesaid parties and witnesses after I have read over to them the present act, have signed it with me, except Jacques Edmond the wife who stated her inability to sign.

Signature of the married parties: Philibert Ducrotoy; Julia Edmond

In the presence of us: [illegible signatures]

[illegible signature]

Officer of the Civil Status of the Island of Seychelles.

Marriage of Philibert Olivier Julefy Ducrotoy from France and Marie Josephe Julia Edmond of Mahe, Seychelles 1891.

SUMMARY OF THE HISTORY OF SEYCHELLES.

"The Seychelles Islands was so named after a Minister of Finance in the reign of Louis XV, Vicomte Moreau de Seychelles. There is some evidence to suggest that they may have known or visited by traders for it is believed that between 200 and 300BC the Polynesians began to move from some place around the Bay of Bengal and followed two main routes. One went to the islands of the coast of Sumatra and thence to the many islands of the Pacific. The other went to the Maldives and Seychelles Islands to Madagascar where their descendants are still living. However, it was also believed that the Seychelles have been visited by the Portuguese during their exploration in the Indian Ocean and the first map showing these islands was dated 1501.

In 1609 a British expedition known as "The Fourth Voyage of the East India Company under the command of Alexander Sharpeigh visited Mahe and the nearby islands. The first landing by the French was made on 22nd November 1742 under the command of Lazare Picault. However, for fifteen years Seychelles remained forgotten and fear that the English were seeking uninhabited islands for bases to harass French shipping, France decided to take possession Seychelles and set up a Stone of Possession on Mahe engraved with the arms of France in 1756.

The first settlement was set up on the islands of St Anne in 1770 under the command of Brayer du Barre. However in 1794 the British captured Seychelles until Independence Seychelles was a British colony with a Governor appointed by Great Britain."

Information sent in a letter from Mr McGaw, Seychelles. 22 October 1982.

Births for Savy Registered in Seychelles 1871 - 1880

Marie Rose Alphonsine Savy, 27 March 1871.
Victorien Ludovic Savy, 3 April 1871.
Joseph Sidney Savy, 25 January 1872.
Marie Aurore Savy, 4 June 1872.
Charles Desire Despilly Savy, 7 October 1872.
Charles Francois Napolean Savy, 15 January 1873.
Charles Furcy Savy, 10 February 1873.
Marie Lucy Savy, 27 February 1873.
Pierre Antoine Savy, 20 May 1874.
Jean Baptiste Olivier Clodoald Savy, 12 June 1874.
George Olivier Gustave Savy, 9 July 1874.
Marie Charlotte Henriette Savy, 4 November 1874.
Marie Elisabeth Louise Savy, 30 November 1874.
Marie Celerine Pelage Savy, 24 December 1874.
Victorien Savy, 20 November 1875.
Raoul Julien Savy, 5 January 1876.
Noel Ronciere Amedee Savy, 22 January 1876.
Louis Emile Charles Savy, 29 May 1876.
Louis Joseph Auguste Savy, 1 February 1877.
Louis Elise Savy, 19 August 1877.
Antoinette Marie Amelie Savy, 15 October 1877.
Henry Leon Dorcy Savy, 20 October 1877.
Charles Joseph Maurice Savy, 1 May 1878.
Francois Mederic Savy, 4 June 1878.
Marie Naomie Savy, 13 January 1879.
Marie Mauricia Julie Savy, 16 June 1879.
Marie Hersilie Savy, 11 September 1879.
Aristide Anthony Savy, 30 October 1879.
Jean Baptiste Emard Savy, 5 March 1880.
Marie Octavie Rosa Savy, 29 July 1880.

FRANCE _ BOURBON, REUNION ISLAND - ST ANNE, SEYCHELLES.

CHARLES SAVY = Blacas
born C1740 at
Reunion.

- **- Savy**
  of St Anne Island.

- **Charles Dorothe Savy = Marie Josephe Jorre de St Jorre, of Isle de France, Mauritius.**
  born 1766 Bourbon.
  dec.4 Sept.1853,11am.

  - Dorothe Francois Marie Fernand Savy. 5 February 1794.
  - Francois Marie Charly Savy, 26 July 1797.
  - Marie Francoise Olivette Claire Savy, 26 September 1799.
  - Marie Thomas Amede Savy, 31 March 1800.
  - Francois Oliver Mellicourt Savy, 11 Oct. 1802. = Louise Antoinette Aricie de Comarmond.
  - Marie-Joseph Francoise Delphine Savy, 22 September 1804.
  - Charle Dorothe Marie Joseph Edouard Savy, 20 August 1806.
  - Charle Joseph Napolean Savy, 20 August 1808.
  - Charle Marie Auguste Savy, 29 July 1810.
  - Marie Olivier Claire Savy, 18 October 1811.
  - Marie Charlotte Elfrida B.Bye Savy, 30 June 1813.
  - Marie Josephe Dorothee Evelina Savy, 5 April 1815.

Note: Francis Savy and Jean B.Argent were witnesses at the death of Charles Dorothy Savy in 1853.

MARRIAGE OF PHILIBERT DUCROTOY and MAUD SAVY - 1924

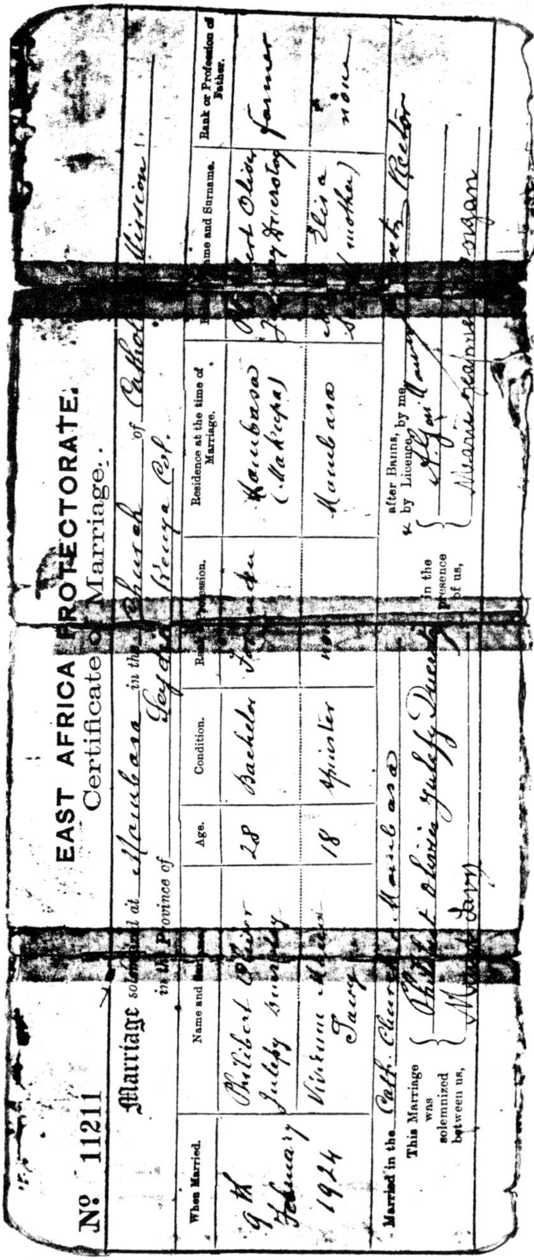

The original marriage certificate of Maud Vivienne Savy of Seychelles and Philibert Olivier Julphy Ducrotoy of Seychelles and France, dated 9th February 1924, Church of the Catholic Mission, Province of Seydie, Mombasa, Kenya. Witnesses to the ceremony: A.Guillaume (Paquerette's husband), and Jeanne Marie Morgan (Philibert's sister). Nancy Savy kindly passed the original to me for safe keeping, which enables me to reproduce it here for everyone's benefit.

REFERENCES TAKEN FROM THE LONDON ARCHIVES RELATING TO FAMILY MEMBERS AND OF GENERAL INTEREST...

Seychelles Returns 1881, Civil Establishments.
Office: Jacques Edmond,*Tide waiter, April 1876 appointed by the Governor. Annual salary 1000 rupees. No house allowance. 31 August 1858 date of first appointment. Period of absence from the Colony -none.
Very Reverend Edmond, Pro-Vicar Apostolic, 22nd December 1981, 5000 Rs.
Circa 1900. On Legistrative Council - official member -Francois Elysee Savy. Re-appointed on 30 December 1904 for three years.
List of Officers: Fanchin Barbe, assistant lamplighter.
Antoine Jorre de St Jorre, 2nd Customs Officer 720Rs.
Edmond Jorre de St Jorre, copiest J.P's Office, Anse Royale.
Dolphin Lalande, assistant messenger, Legal Department.
Mrs James Lebon, Matron of the Prison Department.
Dorset Savy,2nd clerk registry, Legal Department, 610Rs.
Gustave Savy, Sergeant of Police.
Pensions: Dorothe Etienne 180Rs from 1st June 1890, messenger of the Court, retired old age, 70 years.
Cleo Lebon,162Rs, October 1899, Corporal, retired with bodily infirmity, aged 47 years.
Elysee Savy, 30 December 1898 Member of the Legistrative Council.
Dorset H.L.Savy, Junior clerk, 1902.
Antoine Jorre de St Jorre , 3rd clerk, Registry Legal Department 1902.
Sister Gabrielle was the nurse at Victoria Hospital.
Dorset Savy, clerk to Assistant Manager's Office and JP, South Mahe.
Gustave Savy, Sergeant of Police.
1904 Dolphin Lalande, Assistant Messenger Legal Department.
Francois Lalande, Assistant Light Keeper, Denis Island, Port and Marine Department.
1907 Felix Barbe, Inpector of roads.
Emmanuel Ferrari, apprentice, Public Works Department.
Sister Gabrielle, Victoria Hospital.
Antoine Jorre de St Jorre, 2nd Customs Officer.
Edmond Jorre de St Jorre,Copyist, JP's Office, Anse Royale.
Miss Tadela Pothin, female attendant Lunatic Asylum, South Mahe.
1907 Dorset Savy, 1st Clerk Reg. Legal Dept.
Gustave Savy, the gaoler.
1908 Antoine Jorre de St Jorre, Inspector of Distilleries.
1908 Edmond Jorre de St Jorre, clerk, audit office.
Mrs James Lebon, Matron of Prison.

Jacques Edmond was, in all probability, Philibert Ducrotoy's father-in-law, the father of Marie Julia Edmond. It was known by the family that he held a Government post and well educated, as indeed were all the Edmond family.

***

VIGNACOURT FAMILY 1861-1995

- **Joseph Alfred Ducrotoy** 1861 - 1912 born & died Vignacourt
  - **Dorothy Andrea Marie Judith** 1904-1943
    - **Monique** 1926
    - **Yves**
    - **Lillian**
    - **Francoise**
  - **infant** d.1903
  - **Joseph Albert** 1893-1974
  - **Leon** 1889 (twins)
    - **Serge Ducrotoy** 1911 Major in French Army died at War.
    - **Leonel Ducrotoy** died War
  - **Joseph Alfred** 1889-1889 (twins)
- **Philibert Ducrotoy** 1866 - 1924 born Vignacourt, married Seychelles,died Kenya.

Philibert's brother in Vignacourt born 25th October 1861 married Marie Aurelie Angelie Lefebre.Joseph died 3 October 1912 at 9pm. They had five children:

Dorothy Andrea Marie Judith Ducrotoy born 28 January 1904, 9pm,died 23 April 1943. She married Louis Beauval and had four children.

Joseph Albert Ducrotoy born 12 November 1893 died 28 August 1974 sans issue.

Infant Ducrotoy born 10 February 1903.

Leon Ducrotoy born 4 March 1889, a twin,and died 23 January 1940 married Elise Gambier and had:

Serge Ducrotoy born 1911, became Magistrate and Major in French Army.(died War)

Leonel Ducrotoy (died in War)

Monique born 22.1.26 = Serge Deray, and had Coralie 10 December 1966
Judith 18 May 1962
Patrick 1956

Yves married and had four children.

Lillian married Bernard Deray, brother of Serge, and had Muriel Deray.

Francoise married Michel Azais, now living in Toulouse, had two children:
Frank Azais
Sylvie Azais

CHATEAU DU CROTOY.

All those with the name Ducrotoy, and there are few, stem from the original family that migrated from the coast, to the village. The interesting fishing village is named Le Crotoy, hence Du Crotoy (of Crotoy). It is only twenty five miles from Vignacourt where the Ducrotoys finally settled. Anyone studying genealogy will agree with me that the present day Ducrotoy family are not too distantly related.

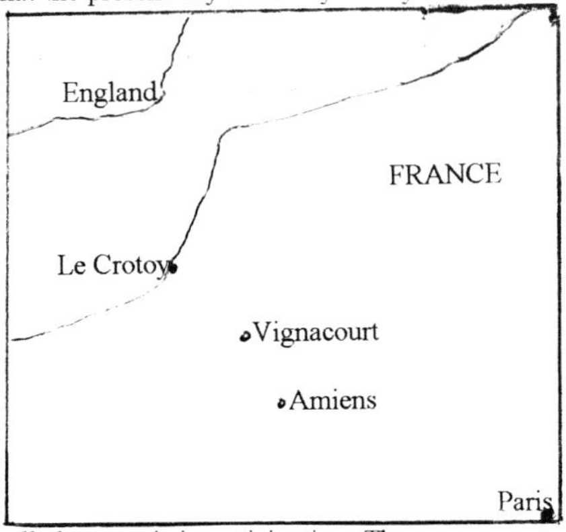

The Crotoy area figured prominently in the past and the Chateau DuCrotoy was a fifteenth century castle. Before the Revolution the chateau belonged to King Louis X1V, Count of Ponthieu and to the Abbey of St Riquier. Joan of Arc was imprisoned at Crotoy. The area was of strategic importance as it commanded the area of St Valerie in the mouth of the Somme.

There is a church at Crotoy, beautifully adorned and the area is a site of many Roman antiquities. On a board outside the remains of the chateau the following is written, "Vestiges du Chateau du Crotoy. Construit vers 1150 par les Comtes De Ponthieu Demanele en 1674".

When we learnt of this chateau in 1975 we travelled to see it in anticipation. There was great disappointment to find a heap of large stone ruins. It was dusk so unable to explore at length but we will return one day...

⚓⚓⚓⚓⚓⚓⚓⚓⚓⚓

The translations on this page relate to the following two certificates:

BIRTH CERTIFICATE 1866 Philibert Olivier Julphy Ducrotoy.

1866 the 13th August at nine hours in the morning, before us, Dubois Alexis, Mayor and Registrar of the Parish of Vignacourt, District of Picquiny, Somme, appeared before us, Celestine Marie Madelaine Godefroy aged 63 years, midwife, resident of Vignacourt, who has assisted at the delivery of the infant undernamed. We declare that the same day of the same month at 8 hours of the morning, the named Irma Thuillier aged 32 years, factory worker, residing at Vignacourt, widow of Joseph Ducrotoy (died at Amiens 18 December 1865) who confined in that district in the house of Olivier Thuillier, labourer, her father, situated Rue de Labas, an infant masculine, which she presents and declares giving the names of Olivier Julphy Philibert, the said declaration and presentation was done in the presence of Pierre Joseph Binet, aged 47 years, panettier, and Theophile Pecourt aged 47 years, both residing at Vignacourt, the first witness announces not able to sign, and called upon the second witness to sign with us preset after reading.

DECLARATION OF JOSEPH DUCROTOY'S DEATH 1865, father of above.

Joseph Ducrotoy, 1865 on 19th December at 1pm, before Pierre Francois Louis Bon Delegue Maire of Amiens. Edward (Edouard) Ducrotoy, 24 years, poultry dealer, living at Vignacourt, Somme brother of deceased, and Louis Ducrotoy aged 53 years poultry dealer living at Amiens, at Hotel de Ville number 27, relative of the deceased, do declare at 7am the death * at Amiens, Rue de Moulin 9, Joseph Ducrotoy aged 31 years poultry dealer born at Vignacourt and living there, spouse of Irma Crognier ** aged 31 years, housewife, son of Florent Ducrotoy aged 57 years, labourer and of Judith Pruvost aged 54 years, labourer, his wife, living at Vignacourt.

* 7am the day before. He died 18.12.1865. ** Irma Crognier(error, she was Thuillier).

BIRTH CERTIFICATE 1866
PHILIBERT OLIVIER JULPHY DUCROTOY.

/JG

Département de la Somme

Arrondissement d'Amiens

CANTON DE PICQUIGNY

REPUBLIQUE FRANÇAISE

MAIRIE DE VIGNACOURT

Téléphone 26

Objet :

Référence :

(A rappeler dans toute correspondance)

COPIE INTEGRALE DE L'ACTE DE NAISSANCE N° 45
de DUCROTOY Olivier Vulphy Philibert

L'an mil huit cent soixante six, le treize Aout, à neuf heures du matin, par devant Nous, DUBOIS Alexis, Maire, Officier de l'Etat civil de la Commune de VIGNACOURT, canton de PICQUIGNY, département de la SOMME, est comparu Célestine Marie Madeleine GODEFROY, âgée de soixante trois ans, sage-femme, demeurant à VIGNACOURT, laquelle a procédé à l'accouchement de l'enfant ci-après nommé et nous a déclaré que le même jour dudit mois, à huit heures du matin, la nommée Irma THUILLIER, âgée de trente deux ans, ouvrière de fabrique, demeurant à VIGNACOURT, veuve du sieur Joseph DUCROTOY, décédé à AMIENS le dix huit Décembre mil huit cent soixante cinq, est accouchée en cette commune, en la maison du sieur Olivier THUILLIER , manouvrier, son père, sise rue de Labas, d'un enfant du sexe masculin qu'elle nous présente et auquel enfant elle a déclaré donner les prénoms de OLIVIER VULPHY PHILIBERT; les dites présentation et déclaration faites en présence de Pierre Joseph BINET, âgé de quarante sept ans, pannetier, et de Théophile PECOURT, âgé de quarante sept ans, ménager, tous deux demeurant à VIGNACOURT, le premier témoin a déclaré ne savoir signer, de ce interpellé, la comparante et le second témoin ont signé avec Nous le présent acte après lecture.

suivent les signatures.

CERTIFIEE EXACTE ET CONFORME LA PRESENTE COPIE

VIGNACOURT LE 6 AVRIL 1976

P.O. LE MAIRE

BLEUET

(SOMME)

DEATH CERTIFICATE OF JOSEPH DUCROTOY 1865
OF AMIENS. (Father of Philibert)

1588
Joseph
Ducrotoy

L'an mil huit cent soixante cinq, le dix Neuf Décembre à une
heure de relevée, par devant nous Pierre François Louis Bon,
Adjoint & délégué du Maire de la Ville d'Amiens, faisant les fonctions
d'officier de l'Etat-Civil; ont comparu Edouard Ducrotoy, âgé de
Vingt quatre ans, marchand de Volailles, demeurant à Vignacourt (Somme)
frère du décédé et Louis Ducrotoy, âgé de Cinquante trois ans, marchand
de volailles, demeurant à Amiens, place de l'hotel de Ville, numéro
Vingt sept, allié au défunt; Lesquels nous ont déclaré que la veille
vers sept heures du matin est décédé à Amiens, Rue du Moulin neuf,
(ce dont nous nous sommes assuré) Joseph Ducrotoy
âgé de trente un ans, marchand de Volailles, né à Vignacourt
(Somme) y demeurant, Epoux de Irma Crognier, âgée
de trente ans, ménagère; fils de Florent Ducrotoy, âgé de
Cinquante sept ans, manouvrier & de Judith Pruvost âgée de
Cinquante quatre ans, manouvrière, son Epouse, demeurant à
Vignacourt; Et après lecture du présent acte, nous l'avons signé seul
les comparants ayant déclaré ne savoir signer.

Bon

MARRIAGE CERTIFICATE OF JOSEPH DUCROTOY and IRMA THUILLIER of France (Philibert's parents). 4th April 1861 at 7am, Civil Ceremony.

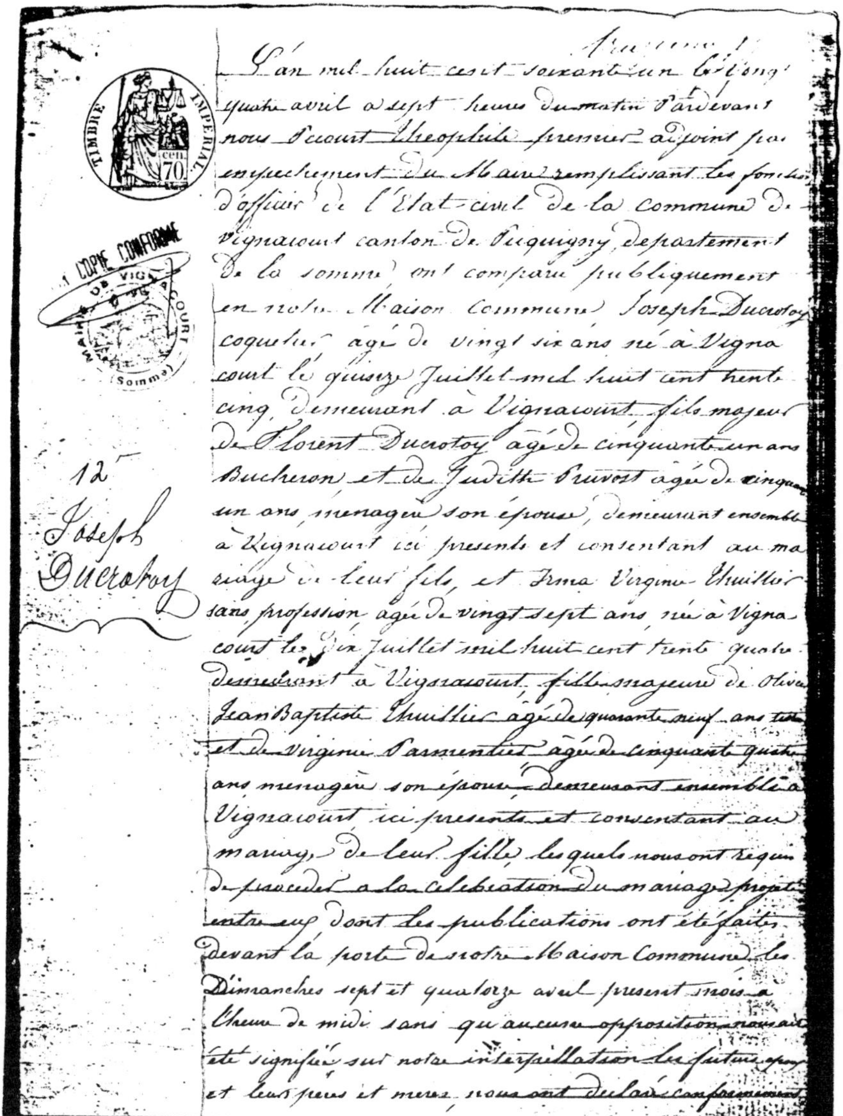

TIMBRE IMPERIAL cen. 70

COPIE CONFORME

MAIRIE DE VIGNACOURT (Somme)

12 Joseph Ducrotoy

L'an mil huit cent soixante un le vingt quatre avril à sept heures du matin pardevant nous Pecourt Theophile premier adjoint par empêchement du Maire remplissant les fonctions d'officier de l'Etat civil de la commune de Vignacourt canton de Picquigny département de la somme ont comparu publiquement en notre Maison Commune Joseph Ducrotoy coquetier âgé de vingt six ans né à Vignacourt le quinze Juillet mil huit cent trente cinq, demeurant à Vignacourt fils majeur de Florent Ducrotoy âgé de cinquante un ans Bucheron et de Judith Prevost âgée de cinquante un ans, menagère son épouse, demeurant ensemble à Vignacourt ici presents et consentant au mariage de leur fils, et Irma Virginie Thuillier sans profession âgée de vingt sept ans, née à Vignacourt le dix Juillet mil huit cent trente quatre demeurant à Vignacourt, fille majeure de Olivier Jean Baptiste Thuillier âgé de quarante neuf ans [illegible] et de Virginie Parmentier âgée de cinquante quatre ans menagère son épouse, demeurant ensemble à Vignacourt ici presents et consentant au mariage de leur fille lesquels nous ont requis de proceder à la celebration du mariage projeté entre eux dont les publications ont été faites devant la porte de notre Maison Commune les Dimanches sept et quatorze avril present mois à l'heure de midi sans qu'aucune opposition nous ait été signifiée sur notre interpellation les futurs époux et leurs peres et meres nous ont declaré conformement

Translation of above: 1861 4th April at 7am in front of us Pecourt Theophile, premier adjoint taking the place of Mayor, Officer of the Civil State Commune of Vignacourt, County of Picquiny, Department of the Somme. Have appeared publicly in our community house (Town Hall), Joseph Ducrotoy (coquetier) age 26, born in Vignacourt 15 July 1835, living in Vignacourt, first son of

Florent Ducrotoy aged 51 years, woodcutter, and Judith Pruvost aged 51 years, housewife of the husband living together at Vignacourt both present and consenting to the marriage of their son and Irma Virginie Thuillier without profession age 27 born in Vignacourt 10 July 1834, living at Vignacourt, eldest daughter of Olivier Jean Baptiste Thuillier age 49 years and Virginie Parmentier age 54 years housewife of her husband living together at Vignacourt, are present and consenting to the marriage of their daughter of whom we are about to carry out the ceremony of the projected marriage between them, the banns have been published in front of the door of the Town Hall (Maison Commune) the Sundays 7th and 14th April this month to the hour of midday

a la loi du dix juillet mil huit cent cinquan
qu'il n'avait pas été fait de contrat de mari
faisant droit aux requisitions des parties et qu
avoir donné lecture des actes de naissance des futu
époux des publications de mariage lesquels son
inscrit sur les registres de cette commune et
du chapitre six du titre du mariage au code
Napoleon nous avons de chacune des parties
l'une après l'autre la declaration qu'elles
veulent se prendre pour mari et femme e
nous avons prononcé au nom de la loi que le
sieur Joseph Ducrotoy et la demoiselle
Irma Virginie Thuillier sont unis par
le mariage de quoi nous avons dressé acte en
presence de Alcide Ducrotoy agé de vingt quatre ans
bucheron frere de l'epoux de Desiré Ducrotoy agé de
cinquante six ans debitant, oncle de l'epoux de Odile
Lecointe agé de vingt trois ans Bourrelier et de clovis
Calippe agé de trente ans, meunier, tous deux amis de
l'epoux tous demeurant a Vignacourt et lecture a
eté donné du present acte aux parties a leurs peres
et meres et aux temoins qui ont declaré ne savoir signer
de ce interpellés excepté Odile Lecointe, clovis calippe, le
pere de l'epouse et l'epoux qui ont signé avec nous.

J Ducrotoy   Thuillier

Lecointe Odile   Calippe Clovis

E Lecomte
adjoint

without objections being presented, the future parents have agreed to the Law of 10th July 1850. We have not read the birth certificates. The publications of marriage that is written in the register of this area end of chapter 6 of the Marriage Act under the code of Napolean. We have on each of the party both declarants that they wish to take each other as husband and wife and we have pronounced in the name of the law that Joseph Ducrotoy and Irma Virginie Thuillier are united by the marriage that we have performed. In the presence of Alcide Ducrotoy, 24 years, wood merchant, and brother of the groom, and also Desire Ducrotoy 56 years being uncle of the groom, and Odile Lecointe aged 23 years, harness maker, saddler, and Clovis Calippe, 30 years, two friends of the bride and all living in Vignacourt; the Act was read to the parties, parents and witnesses who understood and it was read to them, except the bride and groom who has signed with us...Signed JD,IT,OL and CC.

Vignacourt Church.

JORRE DE ST JORRE FAMILY>

Widow Jorre de St Jorre
dec.before 1802

6children

Francois Beatrix Matherine Jorre = Juanne

Marie Joseph Jorre St Jorre = Charles Dorothe Savy

Guillamin Joseph Jorre St Jorre = Marie Ann Michelle

Olivier Ferrant Pitel = Marie Noell Genu

Jeane Francois Marie Jorre = Marie Pitel
Official of National Centre — **b C1754-59**
Marr. 7 November 1774

Marie Corentine Olive Jorre
25 May 1778

Louis Noel Jorre
11 March 1781
dec 25 Dec 1782

Andre Francois Pierre Jorre
7 December 1788

Hilda at Kitale.

Hilda and Rita Ducrotoy as bridesmaid at Therese at the wedding of cousins Therese Ernestine and Tim Morgan.

Albert in Kenya 1958

Catholic Church in Kakamega,Kenya.

Maude Savy-Ducrotoy 1974